PAPE LATYR FAYE

DETERMINANTS OF HOME DELIVERY

PAPE LATYR FAYE

DETERMINANTS OF HOME DELIVERY

Analysis based on a population under demographic and health surveillance

ScienciaScripts

Imprint

Any brand names and product names mentioned in this book are subject to trademark, brand or patent protection and are trademarks or registered trademarks of their respective holders. The use of brand names, product names, common names, trade names, product descriptions etc. even without a particular marking in this work is in no way to be construed to mean that such names may be regarded as unrestricted in respect of trademark and brand protection legislation and could thus be used by anyone.

Cover image: www.ingimage.com

This book is a translation from the original published under ISBN 978-620-6-69524-0.

Publisher:
Sciencia Scripts
is a trademark of
Dodo Books Indian Ocean Ltd. and OmniScriptum S.R.L publishing group

120 High Road, East Finchley, London, N2 9ED, United Kingdom
Str. Armeneasca 28/1, office 1, Chisinau MD-2012, Republic of Moldova, Europe
Printed at: see last page
ISBN: 978-620-7-63404-0

CONTENTS

Summary

In Senegal, according to data from the Enquête Démographique et de Santé version 2019 (EDS - 2019) published by the Agence Nationale de la Statistique et de la Démographie (ANSD), the proportion of women who used a health institution during childbirth rose from 62% in 2005 to 80% in 2019. In the Niakhar observatory, use of healthcare during pregnancy and childbirth has evolved fairly slowly over time. In the Niakhar SSDS, the rate of healthcare use during childbirth was 11% between 1988 and 1992; 16% in 1998 and 2002; 34% in 2008 and 2012, and 57% in 2017, despite an acceptable supply of healthcare, but with a highly differentiated distribution across geographical units. Indeed, the rate of home births in the Niakhar observatory is lower than the national rate. This study is cross-sectional, descriptive and explanatory, with the aim of identifying the factors that explain non-use of medical assistance in childbirth in the Niakhar observatory area. Retrospective data from the Niakhar database were used to calculate the proportions of home births from 1983 to 2020. The chi-square test and binary logistic regression were respectively used at the descriptive and explanatory levels on qualitative data from a field survey of women at the SSDS - de Niakhar, conducted by myself as part of this work. The analysis was carried out on the following variables: parity achieved, marital status, number of ANCs during the last pregnancy, level of education, woman's caste, woman's age, woman's economic situation.

<u>Key words</u>: Medical assistance; home delivery; women; Niakhar observatory; Senegal

INTRODUCTI ON

While motherhood is often a positive and satisfying experience, it is unfortunately, for too many women, synonymous with suffering, illness and even death (WHO, 2018). Women lose their lives as a result of complications arising during or after pregnancy and childbirth. Most of these complications appear during pregnancy and could be avoided or treated. The main complications, accounting for 75% of maternal deaths, include: severe haemorrhage (especially after delivery), infections (usually post-delivery), hypertension during pregnancy (pre-eclampsia and eclampsia), delivery complications and unsafe abortion.

However, the illustration of this phenomenon follows a spatiotemporal logic that manifests itself very differently in different geographical areas. "In industrialized countries, almost all women benefit from assisted care and delivery; however, one in twenty-six (26) women in sub-Saharan Africa, compared with one in 7,300 in these countries, is at risk of losing her life in the course of her lifetime" (M. Munyemana et al, 2010, page 21). In the least developed countries, more than half a million mothers die every year as a result of childbirth and pregnancy. Eighty-nine percent of these maternal deaths occur in less developed regions, most of them due to a lack of adequate care at the time of delivery (Ransom, 2002). It has been shown that it is possible to reduce the risks associated with motherhood for all these women. This is why maternal mortality has become a major concern for the WHO, development actors, and is also at the heart of the Sustainable Development Goals (SDGs) in part 3.1 of goal number three (3).

With this in mind, the Senegalese government has drawn up its National Health and Social Development Plan (PNDSS) for the 2019-2028 period, extending the 2009-2018 plan. This plan aims for "a Senegal where quality curative, preventive and promotional care is accessible to all social strata of the population without any form of exclusion, and where an economically and socially productive level of health is guaranteed" (MSAS, 2009). The introduction of the universal health coverage program guarantees the population access to a minimum package of care, including free caesarean sections, free care for children aged 0 to 5 and free haemodialysis, especially for those living in rural areas. The objectives of this policy are to reduce maternal mortality, reduce infant and child mortality, and control fertility.

An analysis of the data available in the 2019 Demographic and Health Survey (EDS-2019) report published by the Agence Nationale de la Statistique et de la Démographie (ANSD) reveals a steady trend since 2005 in the proportion of women who used a health institution during childbirth, rising from 62% to 80% in 2019. However, a closer look at these statistics reveals discontinuities in the rate of institutional deliveries between urban areas (96%) and rural areas (76%). In the area of interest of this research, the Niakhar population and health observatory (or SSDS - Système de surveillance démographique et de santé de Niakhar), the oldest SSDS in sub-Saharan Africa, has seen a steady evolution in the use of health facilities during pregnancy and childbirth since 1983, rising from

11% in 1988-1992, to 16% in 1998-2002, to 34% in 2008-2012, then to 57% in 2017, albeit with geographical heterogeneity (Delaunay, 2017).

According to Valérie Delaunay (2017, page: 71) in the Niakhar SSDS, "generally speaking, young women use health facilities more frequently than older women, whether for prenatal consultations or childbirth. The mother's age is a socio-demographic determinant of the care and medical follow-up of women's pregnancies, but it is not the only one. We have found that women's level of education also plays a role in the medical follow-up of pregnancies. The better educated a woman is, the more likely she is to benefit from prenatal care and medicalized deliveries".

However, these determinants are not the only ones to influence the use or non-use of a health institution during pregnancy and childbirth. The available literature indicates that parity achieved, marital status, number of prenatal consultations, woman's caste, woman's economic situation, woman's age and woman's level of education also play a role in whether or not a health institution is used during pregnancy and childbirth.

Against this backdrop of improving healthcare provision and recourse to institutional childbirth, our focus will be on home births, with a view to providing some answers as to how these behaviours are maintained.

I.   Subject of research:

The research questions guiding our work are as follows:

1) Against a backdrop of ever-increasing but heterogeneous healthcare provision in the Niakhar SSDS, is the distribution of home births differentiated across geographical units?
2) In a rapidly changing rural context, what are the determinants influencing healthcare use during pregnancy and childbirth in the Niakhar observatory?

General objective: The aim of this research work is to analyze the spatial distribution of home births in the Niakhar opulation observatory and to study the factors contributing to explain this phenomenon.

The specific objectives of this research work are as follows:

-   To analyze the spatiotemporal distribution of deliveries outside health institutions in the various villages of the Niakhar SSDS.
-   Determine the profile of women not assisted at birth by skilled health personnel and identify the main determinants influencing home birth.

The general hypothesis of this research work is to analyze that home birth results from a set of constraints or obstacles linked to the socio-cultural or socio-demographic characteristics of women, the socio-economic characteristics of households, and the inaccessibility and lack of quality of obstetric services.

Specific research ***hypotheses***: it follows from our general hypothesis that home birth is specifically determined by:

-age at delivery   - parity achieved

-women's marital status   -women's caste

-level of education   -women's economic situation

-the number of CPNs.

This research work is structured in three (3) parts: the first part consists of a bibliographical synthesis; the second part is devoted to a description of the material and methods used; finally, the third part presents the results, discussions and draws conclusions."

II.   Defining concepts.

We begin by defining some of the terms that will be used in this document to help understand the study of the use of health care facilities during pregnancy and childbirth. In the second part of this chapter, we will then look at what the literature has to say about home births.

-    Health :

In the preamble to the 1946 WHO constitution, health is defined as "a state of complete physical, mental and social well-being and not merely the absence of disease or infirmity. The enjoyment of the highest attainable standard of health is one of the fundamental rights of every human being without distinction of race, religion, political belief, economic or social condition". Taking into account the socio-spatial and economic dimension is of paramount importance in improving the health of populations. Factors contributing to health include access to health services, housing, food, education, work, a healthy environment, etc.

- Maternal and neonatal mortality :

Maternal mortality is defined as a phenomenon associated with pregnancy and childbirth. According to WHO (2016), it is "the death of a woman occurring during pregnancy or within 42 days of its termination, regardless of duration or location, from any cause determined or aggravated by pregnancy or the care it prompted, but neither incidental nor accidental".

Neonatal mortality includes children born alive but dying between birth and the 28th day of life. A distinction is made between early neonatal mortality, for deaths occurring within the first week, and late neonatal mortality, for those occurring within the following three weeks.

- Use of healthcare :

This is the use of the healthcare system. The use of healthcare must be distinguished from access to healthcare, which merely reflects a material or regulatory capacity to use healthcare services. The use of healthcare depends on a set of interrelated economic, social and cultural factors. The most important issues concern access to specialized and costly care, prevention and the time it takes to access care.

The use of healthcare is then determined by several factors, such as income, level of education, social background, place of residence, gender, age... The use of healthcare depends on a set of factors whose respective influences are difficult to distinguish (P. Lombrail; J. Pascal, 2005).

III. Literature review

The literature on maternal health care utilization clearly shows that the use of health care facilities during pregnancy and childbirth depends largely on the sociodemographic and economic profiles of parturients, but also on the institutional environment in which they are based. The dynamics of home births and the factors that explain them are well documented. However, the factors that influence whether or not to use a health institution during pregnancy and childbirth are not perceived in the same way by the authors. According to Ba Gning and Sandberg (2018), "the main determinants of health care utilization are cost, availability and quality of services, social structure, community health beliefs and cultural perspectives, which also have a determining influence on health decisions and behaviors". Kroeger (1983), in a study of anthropological and socio-medical health research in developing countries, argues that "patterns of health service utilization by communities are the product of predisposing individual and economic factors, as well as of a supply of care characterized by its quality and accessibility".

On the other hand, Beninguisse et al (2003), in a study carried out in Cameroon, question the relevance of the geographical accessibility explanation. They show that it is the mismatch between supply and women's cultural expectations and preferences that explains their decision not to visit health facilities and to give birth at home.

Consequently, the factors influencing parturient home birth can be classified into two broad categories: predisposing factors (such as the woman's age, parity, marital status, spouse's profession, woman's education or literacy, household standard of living, health capital, woman's religion, etc.) and facilitating factors (such as accessibility to health services, availability of services and quality of obstetric care, etc.).) and facilitating factors (such as accessibility to health services, availability of services and quality of obstetric care, etc.). These categories are also known as "predisposing factors" and "enabling factors" (Andersen and Newman, 1972).

III.1 Factors related to the use of obstetric care or predisposing factors

The factors mentioned above as predisposing factors are not exhaustive. The choice of variables studied under these factors is linked to the context of the study area and the particularities represented by our research field.

Socioeconomic characteristics

- Household socio-economic level

According to Zoungrana (1993) and Beninguisse (2001), the socio-economic level of households is a significant determinant of healthcare use in both developed and developing countries. According to Lavy and Quigley (1993), and Dor and Van der Gaag (1988) (cited by Gnanderman Sirpe, 2011), income is the main determinant of the use of obstetric health services. Similarly, Gertler and Van der Gaag (1990) show that individuals living in relatively high-income households have a higher probability of seeking health care than those in poor households. Beninguisse et al (2007), in a research study on discontinuity of obstetric care in sub-Saharan Africa, note that "discontinuity of obstetric care is significantly more frequent among women residing in households with a low standard of living, and its prevalence decreases significantly as the household standard of living increases".

- Women's economic activity :

A number of studies link women's paid work to better use of health facilities during pregnancy and childbirth, and consequently to better maternal health. The financial autonomy of parturients significantly increases their ability to use health facilities when needed (Kaboré, 2005). This relationship was highlighted in a study carried out in Kampala, Uganda (Nankwangua, 2004). For some authors, women's economic activity also enables them to accumulate social capital, enabling them to adopt new health behaviours to the detriment of traditional health habits (Beninguisse, 2003). However, the relationship between a woman's economic activity and her use of health care is not always easy to establish and therefore remains relative, as some economically active women have cited lack of time as a reason for not using maternal health care (Wong et al., 1987; Macdonald, 1988; Zoungrana, 1993).

Socio-demographic characteristics

- Age and parity

Age and parity are highly significant variables in the study of health behavior, particularly maternal health. In many cases, there is a correlation between age, parity and use of obstetric care (Rakotondrabe, 2001; Sala-Diakanda, 1999; Zoungrana, 1993; Beninguisse, 2003).

In a study conducted by N. Nkurunziza (2015) in rural Burundi, the qualitative survey showed that: "since the daily running of the household relies on the woman, some multiparous women take the risk of waiting to avoid spending too much time in a health facility. Others give up going, as they can't take time away from their household to look after their children during consultations". Age and parity, in various forms, are obstacles to the use of obstetric care; for example, some multiparous women of advanced age find it very inconvenient to use health facilities where they would be cared for by very young medical staff. Such cases are also observed in the work of Rakotondrabe (2001) on the contribution of gender to the explanation of children's health in Madagascar, where "the risk of a woman not seeking care during pregnancy is halved when she is under 35". This observation was also made by Diallo and colleagues in Guinea. Similarly, a study in Canada showed that women in older age groups and multiparous women were proportionately more likely to give birth outside the hospital setting. However, the use of health care facilities by primiparous women is well established, since according to De Sousa (1995), "the choice of modern hospital assistance is frequent in first deliveries. We can therefore assume that greater care is taken during first deliveries". However, the frequency with which first pregnancies are attended decreases as women gain experience of maternity (Masuy-stroobant, 1996, cited by Beninguisse, 2001). Similar results appear in various studies, including those carried out in Utter Pradesh, India (Stephenson and Ong Tsui, 2002), Guatemala (Pebley et al, 1996), Bobo Dioulasso, Burkina Faso (Baya, 1999) and Kenya (Ochako et al, 2011).

- Health experience :

The term "health capital" here encompasses all complications during pregnancy and childbirth: caesarean section, miscarriage, stillbirth, etc., which may influence the women's eventual use of healthcare (Beninguisse, 2007).

According to Stephenson and Ong Tsui (2002, p136), quoted by N. Nkurunziza (2015), their research in India has shown that use of a health facility during pregnancy and childbirth is positively influenced by the experience of losing a child. According to Beninguisse (2003), "this surveillance does not really respond to a preventive logic, but probably corresponds to a fear of recurrence of the antecedents or [to a] punctual therapeutic response". In N. Nkurunziza's (2015) work in Burundi, it is the experience of the previous pregnancy and delivery that guides their choices. The occurrence of complications during pregnancy is widely documented in the scientific literature as a motivation for parturient women to seek health care (Chang, 1980, cited by Zoungrana, 1993).

Socio-cultural factors

- Performances :

In the study of therapeutic itineraries, social representations are insufficiently documented. Each society has its own way of perceiving and understanding the social facts with which it is confronted on a daily basis. According to Denise Jodelet (1989), "social representation is a form of knowledge that is socially elaborated and shared, has a practical purpose and contributes to the construction of a common reality for a given social group". For Jean-Claude Abric (2005), "social representation is both the product and the process of a mental activity by which an individual or a group reconstructs the reality confronting it and attributes a specific meaning to it". Although the literature on the reasons influencing the use of healthcare during pregnancy and childbirth is abundant, the analysis of social representations is not very present in the scientific literature. Most of what is available in the literature concerns individual reasons, such as risk perception (Galotti et al., 2000; Lyerly et al., 2007). For example, a study carried out in the Banfang health district in Cameroon (MankolloBassong Oy et al., 2020) on social knowledge and representations of post-sanitary follow-up established that "traditional practices after childbirth are of great importance in our society". This is corroborated by the study conducted by Olivier et al. (1999) cited by Nkurunziza (2015); although these practices vary, they have a significant impact on postnatal consultation attendance

- Religion:

The research available in the literature recognizes religion as a relevant variable, influencing women's perceptions, attitudes and behaviors towards medical use and care during pregnancy (Sidibé, 2019; Kabakian-Khasholian et al., 2000). Religion, through its derived myths, beliefs, norms, values and practices, is seen as a system of practices and beliefs in use within a group or community. According to G. Charles (2014), "over the millennia, each society has developed, through the observation of nature and the creation of myths and religions, a set of protective rites and prohibitions aimed at protecting mothers and their fetuses". Certain similarities are observed, such as the pregnant woman's specific relationship with the spirits and the symbolism surrounding the placenta and umbilical cord. However, even if birth in a health facility is often reassuring because of the security it brings, it can also be unpleasant because of the impossibility of respecting traditions. In Africa, several studies have shown that some religions are more conducive to obstetric care than others. For example, according to Kaboré et al (2005), the Christian religion has a positive influence on the use of modern obstetric care, due to its openness, adaptation and promotion of Western culture, knowledge, technology and medicine, unlike other religions. However, work by Stock (1983) observed that Muslims make little or insufficient use of the health facilities provided by Catholic missions.

- Education and/or literacy:

The link between education and health behavior is well documented and represents an essential variable in the study of population health behaviors (Yanagisawa et al., 2006; Madagi et al., 2007; Singh et al., 2012). Education refers to the process of acquiring knowledge, skills and know-how within an educational system structured and organized according to Western standards. However, the measurement of education varies in studies, with some researchers referring to the number of years spent in the education system, others to the highest level of study (Akoto, 1993). Populations adopt different health behaviors according to their level of education. A higher level of education is often associated with a greater likelihood of using obstetric health care. For example, according to Vallin et al (2002), "knowledge is one of the first assets likely to encourage the adoption of health-promoting behaviours". Education is often synonymous with modernity and openness to the Western world. Beninguisse (2003), quoted by Tchango Ngale Georges Alain (2015), has shown that knowledge acquired at school is essential for integrating women into the modern medical system, enabling them to rationally assess emergency health situations. Thus, education provides access to information and awareness, enabling a better perception of the benefits of using healthcare. For example, a study on the use of maternal health services in Bamako, Mali (Zoungrana, 1993, cited by Sidibé, 2019) showed that uneducated women were less likely to use health facilities compared with educated women, whose use increased with duration or level of education. However, there is no consensus among researchers as to the positive impact of education on health behavior. Some believe that education makes it easier for women to get a job, and therefore to have the economic means to cover the costs of pregnancy and childbirth. For other researchers, education is synonymous with refinement and conformity to Western standards, where health is a major concern (Rakotondrabe, 2004; Akoto, 1993). According to Jaffre and Prual (1993) and Mebtoul (1993) (cited by Sidibé, 2019), the welcome given to parturients depended largely on their level of education, as medical staff classified women according to their cultural expectations.

- The number of prenatal consultations (CPN)

According to Tchango Ngale Georges Alain, (2015) in a research study conducted in Cameroon "the number of ANC performed is the variable that contributes most to explaining the occurrence of home birth for each level of analysis. The relative risk of home birth varies inversely with the number of ANC performed". The correlation between home birth and prenatal consultation is thus established. Women who had not attended any prenatal ANC consultations were at greater risk of having a home birth than other parturients who had attended at least one ANC consultation. In the same study carried out in Cameroon, "women who have not undergone ANC expose themselves to a loss of vital information on this subject, and run the risk of subsequently adopting behaviours deemed inappropriate. In these terms, it is understandable that women who did not undergo ANC during their last pregnancy are more likely to give birth at home than those who did". However, in Cameroon, up to 85% of women had received antenatal care at least once during their last pregnancy, yet for many of them, births still took place at home in the absence of qualified personnel. This suggests that there are socio-cultural and economic reasons for women to give birth at home, whether they have had at least one ANC or not".

These results tie in well with those of Diallo et al (1999) in a study carried out in Guinea, where a strong mismatch between antenatal consultation and assisted deliveries was noted, resulting in very high rates of home deliveries.

- Women's marital status

Among the significant variables influencing the choice of birthplace, women's marital status figures prominently. This is defined as a person's legal or social marital status: single, married, widowed, divorced, common-law, cohabiting, etc. According to Ouattara Kalilou (2019), in a study carried out in the village of Namassi (north-east Côte d'Ivoire), where 70% of the women questioned were in a couple, it was shown that these women are generally under the shadow of their husbands, who in most cases take decisions for them. According to him, "this situation is like a social determinism that has an impact on women's behavior when it comes to managing pregnancy and preparing for childbirth".

However, this variable is also discussed and is the subject of much controversy as to the most representative marital status between married and unmarried, because according to Coulibaly Salimata Kane (2020) contrary to the results of Ouattara Kalilou (2019) unmarried women were the most represented among women giving birth at home. And these results concur with those of Sangho O. et al. On the other hand, they were also contrary to those of Sidibé A. and Keita, who found that married women were the most represented.

- Spouse's occupation

The influence of the husband's profession on the choice of birthplace is very significant in most of the literature (S. H. Idriss et al, 2006; Y. Kamaté, 2019; Diallo Nanténin Kaba Diakité, 2014). According to Joseph Bénie Bi Vroh et al, (2009) in a study carried out on the prevalence and determinants of home births in two precarious neighborhoods in the commune of Yopougon in Abidjan, Côte d'Ivoire, the results showed that out of 236 women surveyed (41.9%) had a spouse who received a daily income, and consequently the Chi test indicated that there was a significant link between the spouse's type of income and the place of birth. These results also appear in the work of (Coulibaly Salimata Kane, 2020).

III.2 Factors affecting the availability of obstetric care or facilitating factors t

In the scientific literature on the use of health facilities during pregnancy and childbirth, authors focus primarily on the demand for care. However, parturients may be confronted with harsh conditions such as long distances to travel between health centers and places of residence, interminable queues on arrival, incompetent medical staff, insufficient sanitary equipment, and a lack of means for emergency evacuation, which may limit their use of a health institution.

- Geographical accessibility of health services :

Geographical accessibility is defined as the availability of healthcare services, measured by the distance or density of healthcare structures or professionals in a given territory. According to Picheral (2001), in health geography, geographical accessibility represents the material capacity to access health services. In the scientific literature, most research carried out in sub-Saharan Africa on home births has established that geographical accessibility is a significant determinant of the use of health facilities during pregnancy and childbirth (Acharya and Cleland, 2000; Magadi, Madise and Rodrigues, 2000; Raghupathy, 1996). In Guinea Conakry, distance between health facilities and place of residence was the most frequently cited reason for home delivery, according to Diallo et al (1999). These results concur with those of Beninguisse et al (2001), who found a significant decrease in the use of health institutions during pregnancy and childbirth in urban and rural areas of Cameroon with increasing distance from the health center. A study carried out in the Nyaruguru district of Rwanda's southern province by M. Munyemana et al (2010) came to similar conclusions, as did the findings of Joseph Bénie Bi Vroh et al (2009).

- <u>Affordability of healthcare services :</u>

The high cost of healthcare services is widely recognized as a factor influencing the use of maternal healthcare in the scientific literature. The results of research carried out in Maroua by Eloundou Messi et al (2017) showed that women sometimes gave birth at home for lack of choice due to the very high cost of obstetric care, which exceeded their financial capabilities. Numerous studies in sub-Saharan Africa have established a negative link between the high cost of health services and home births, as well as the social consequences associated with this financial burden (Gruénais & Ouattara, 2006; Storeng et al, 2008; Ensor & Cooper, 2004; Borghi et al, 2006; Fabienne Richard et al, 2008).

- Quality of obstetric care:

According to WHO (2019), quality of care refers to the degree to which health services for individuals and populations increase the likelihood of achieving desired health outcomes consistent with current professional knowledge. For example, according to the results of a qualitative analysis carried out in Cameroon on cultural accessibility as a quality requirement for obstetric care and services in Africa (Beninguisse et al, 2004), women have quality expectations regarding certain aspects of maternity care, including adequate management of obstetric conditions and complications, care that is easily accessible to all and at a lower cost, respect for discretion during consultations, preservation of their privacy, and the inclusion of traditional care that they believe in because of its effectiveness (such as massage). The same results also show that women deplore the long queues, crowded living rooms, family planning orientation, limited presence of the family circle, and inability to protect mother and child from evil forces. This dissatisfaction with the quality of the care system is also cited as a reason for non-use of maternal health care in the work of O. Kalilou (2019).

Analysis of the elements of the literature review enabled us to construct a conceptual scheme of the variables relevant to the study of home birth.

Figure 1Conceptual diagram of home birth factors according to the literature review

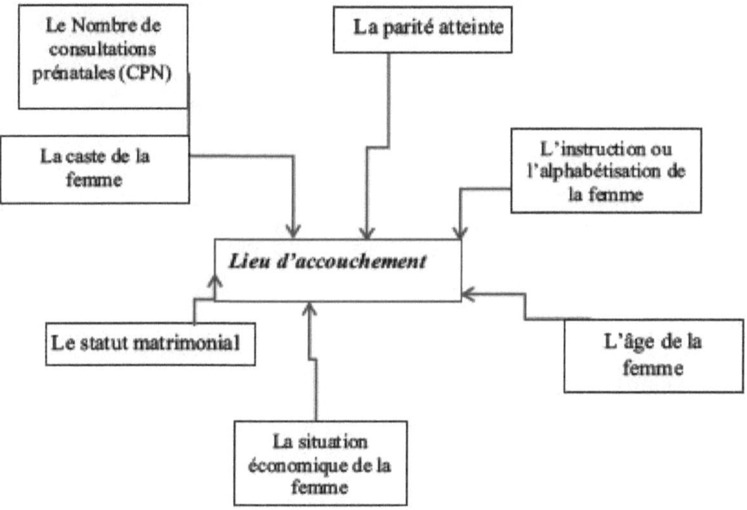

Partial conclusion

Evidence from the literature shows that maternal health care utilization depends on several complementary variables that are relevant to study. However, the existing literature lacks a geographical approach. The work available is carried out by health economists, epidemiologists, demographers, public health specialists and others. This is why it is interesting to develop a geographical approach to this research theme.

IV.   Research methodology

IV.1 Study framework

The Population and Health Observatory, also known as the Demographic and Health Monitoring System (SSDS), brings together a series of field activities and IT procedures to manage the longitudinal monitoring of specific entities such as individuals, households and residential units, as well as all associated demographic and health data within a clearly defined geographical area. The system begins with an initial census to define and register the reference population. Thereafter, regular surveys are carried out to collect data at specific intervals, making it possible to identify new inhabitants, households and residential units, as well as to update the key variables and characteristics of entities already enumerated. The central system monitors demographic dynamics by conventionally collecting and processing data on births, deaths and migrations, the only demographic events that can alter the initial size of the resident population. Thus, the Niakhar Observatory was created in a context where data were scarce after independence, with the aim of having a solid base of accurate information on demography, health and reliable statistics to effectively support development strategies. Pierre Cantrelle, a medical demographer at the Office de la Recherche Scientifique et Technique d'Outre-Mer (ORSTOM), now the Institut de Recherche pour le Développement (IRD), established a multidisciplinary research platform in Niakhar in 1962 to describe the state of the population, monitor its evolution and serve as a model for development aid, beyond geographical, socio-economic and politico-historical boundaries (Pierre Cantrelle, 2018).

The current study area includes eight villages in the Ngayokhème zone, under continuous demographic monitoring since 1963, as well as twenty-two other villages first surveyed in 1983, part of the Niakhar arrondissements and the Diarrère rural community (Cheikh Sokhna et al, 2018). The Niakhar Observatory is located in the department of Fatick, in the heart of the Senegalese groundnut basin. It is fifteen kilometers long and fifteen kilometers wide, covering an area of around 230 km2. The Sahelian region is mainly made up of baobab, rônier and tamarind trees. At present, with the government's loss of interest in groundnut production, this is gradually being replaced by millet as the main food resource. However, farmers face numerous challenges due to demographic pressure and climate change. Livestock farming (sheep, goats, cattle and even pigs) is also an important activity in the area. Today, the area is home to over 50,000 inhabitants, divided into hamlets and concessions with an average population of sixteen. The population of the Niakhar SSDS is young: 58% are under twenty. Muslims make up 70% of the population and Christians 25%. While half of the men have had some schooling, however brief, over 75% of the women have received no education whatsoever. According to Valérie Delaunay et al (2017), between 2003 and 2014, in the Niakhar area, 16% of households were considered poor according to subjective, monetary and non-monetary measures. Being a high-density rural area (over 130 inhabitants per km2), seasonal urban migration has become widespread, especially among young people. On January 1, 2014, among residents, 90% of men aged 30 to 34 and 70% of women aged 20 to 24 had already made a temporary migration for work (Valérie Delaunay, 2018).

Figure 2SSDSS - Niakhar location map.

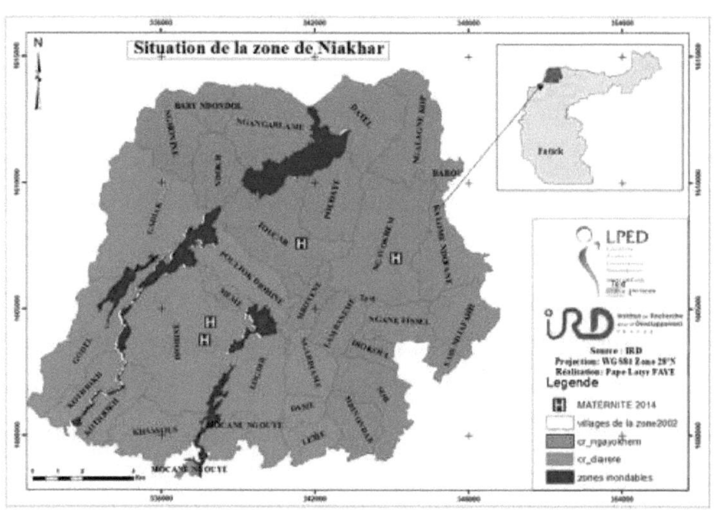

- Presentation of the SSDS - Niakhar care offer.

Four dispensaries cover the Niakhar study area, located respectively in the villages of Toucar, Diohine and Ngayokhème. The Toucar health post, established in 1953, is the oldest in the Niakhar zone, serving ten villages: Mboyène, Ndokh, Ngangarlam, Ngonine, Bari Ndondol, Datel, Lambanème, Poudaye, Toucar and Poultock Diohine. Ngayokhème, established in 1983, covers nine villages: Ngalagne Kop, Mbinondar, Kalome Ndofane, Diokoul, Ngayokhème, Sass Ndiafadj, Sob, Ngane Fissel and Darou. As for Diohine, it has two health structures: a private Catholic health post created in 1957 by the Catholic congregation "les Filles du Saint-Cœur de Marie", and a public health post opened in 2014, including a maternity ward. These two health posts cover some twenty villages, twelve of which are included in the Niakhar observatory zone: Diohine, Dame, Gadiack, Godel, Khassous, Kotiokh, Lème, Logdir, Même, Mokane Gouye, Ngardiam and Poultock Diohine. However, the catchment areas of these health institutions remain flexible, as populations go beyond the nearest health posts to use other establishments for specific healthcare needs. These health posts are supported by health huts, although most of them are not operational (see tables 1 and 2).

Table 1Health infrastructure in the SSDS - de Niakhar 2018.

Ngayokhème* health post (built in 1983 and renovated in 2008)	⟶	Case de Sass (2013) Case de Sob (2010) Case de Kalom Ndofane (2010) Case de Diokoul (2010)
Toucar* health post (built in 1953 and renovated in 2014)	⟶	Case de Ndokh (2017) Case de Ngonine (2003) Case de **Poudaye*** (2002) Case de **Poultock*** (2007)
Diohine* public health post (built in 2014)	⟶	Case de **Gadiack*** (2003)
Diohine* private Catholic health post (built in 1957 and renovated in 2009)	⟶	Case de Kothiokh (2018) Case de Godel (2012)

*** Non-functional health hut / * Health post

Data source: (Hamidou Diallo and Maria-Belén Ojeda DevObs, 2018)

Table 2Niakhar SSDS health post staff.

Health post	Ngayokhème	Toucar	Diohine	Diohine Catholique
Head nurse (ICP)	1	1	1	1
Midwife	1	1	1	1
Community health worker (CHW)	1	1	2	2
Matron	2	4	1	1
Relay (Communication and information dissemination)	3	5	-	-
Pharmacy depository	1	1	1	1

Data source: (Hamidou Diallo and Maria-Belén Ojeda DevObs, 2018)

Data sources

This research is based on two distinct data sources. Firstly, we used data from the longitudinal follow-up of the Niakhar Population and Health Observatory over the period 1983-2020. The choice of 1983 corresponds to the start of observation of the 30 villages making up the current observatory. Each year, all the geographical units of the Niakhar SSDS are surveyed. These surveys gather information on births, deaths, marriages, migration, education levels, place of birth, as well as health data such as vaccinations and causes of death. Demographic trends, such as the decline in maternal and neonatal mortality, or the drop in fertility, are also measured. It is important to emphasize that the data in the Niakhar SSDS database are of rare precision, covering all spatial units in the area.

In addition, the second part of this research work is based on qualitative data from a field survey of women at the Niakhar SSDS, conducted specifically for this work. These data were collected through interviews based on a pre-established interview guide. A total of 46 interviews were conducted, covering a sample of 218 births.

IV.2 Sampling technique

As part of this study, both quantitative and qualitative data were collected in the field. Data collection took place in the village of Mboyéne, a village in the Niakhar observatory chosen for its particularity: it has the lowest proportions of deliveries in health facilities in the whole Niakhar area, despite its proximity to the four (4) health posts serving the region.

Interviews with participants were based on an interview guide and followed the saturation rule. In qualitative research, saturation occurs when the data collected no longer reveals any significant new information. At this point, additional data collection becomes redundant, since the information gathered is already known. Reaching saturation confers a solid basis for generalization, playing a role similar to that of representativeness in questionnaire surveys (Mucchuelli, 1991).

IV.3 Choice of analysis unit

 For research into the health management of childbirth, two types of unit of analysis are possible: births or women (Beninguisse, 2003).

IV.4. Birth-based approach

This approach involves creating an analysis file of births occurring during the study period, starting at the reference date. This method of data collection is more appropriate for studies of children's health. It offers the advantage of a large sample size, thus limiting random errors due to a low number of events. The analysis of behaviours associated with medical assistance during childbirth, based on this birth-centred approach, makes it possible to identify births presenting potential risks and to understand the child's health needs, crucial information for health programs (Beninguisse, 2003). However, this approach has the disadvantage of assuming a degree of independence between women's births. Indeed, recourse to medical assistance during one birth does not necessarily guarantee the same behavior for subsequent births to the same woman (Beninguisse, 2003). In this case, 'the births of the same woman are not statistically independent in terms of participation in obstetric services' (Beninguisse, 2003: p110).

IV.5. Women-based or "women-centred" approach

The woman-centred approach adopted in this study aims to identify at-risk groups, i.e. women who are likely to experience pregnancy complications due to inappropriate health behaviours (Beninguisse, 2003). However, this method has the disadvantage of involving a smaller number of births than the total number of births approach.

IV.6. Interviews

We conducted semi-structured interviews with the aim of identifying the determinants associated with home births. We surveyed 46 women residing in the Niakhar SSDS, who had experienced a total of 218 births. In addition, we had the opportunity to talk to head nurses (ICP) and midwives to assess the maternal health situation, particularly in the Toucar health post, responsible for the village of Mboyéne. The interviews were then transcribed and analyzed using the discourse analysis method, a multi-disciplinary qualitative approach enabling a precise study of discourse.

IV.7. Statistical methods for data analysis.

Retrospective data on home deliveries at SSDS - Niakhar.

Retrospective data on home deliveries from 1983 to 2020, classified by village and district, were extracted from the Niakhar database. They were entered and analyzed using XLSTAT data analysis software. Proportions were calculated to assess the rates of home births in all the villages and neighborhoods of the Niakhar observatory over a diachronic period, highlighting inequalities in the distribution of these births. To better visualize the evolution of this phenomenon, five (5) year intervals were defined (1983-1987, 1988-1992, 1993-1997, 1998-2002, 2003-2007,

2008-2012, 2013-2017), with individual observations for the years 2017, 2018, 2019 and 2020. These data were then linked to the Niakhar SSDS geospatial database for mapping using ArcGIS version 10.4 mapping software, respecting the defined time intervals as well as the different observation scales (village, neighborhood).

- Quantitative data collected during the qualitative survey of women in Mboyéne Village.

It will be structured mainly around two levels of analysis:

- *Descriptive analysis*

We will opt for a bivariate analysis to examine the correlation between non-use of health facilities during pregnancy and childbirth and the explanatory variables. This analysis will be carried out using the chi-square statistical test (chi2, X^2) with STATA version 16 software. The significance threshold for this study is 5%.

Explanatory analysis

The aim of this work is to identify the variables influencing home births in the Niakhar SSDS. Given the binary nature of our dependent variable, we will opt for the logistic regression method. This method is used to estimate event probabilities and establish relationships between characteristics and probabilities of specific outcomes.

In this research, the logistic regression method will be used to explain the behaviour of a qualitative dependent variable using one or more explanatory variables. The dependent variable, representing place of birth, takes the value 0 for institutional births and 1 for home births. The model will predict the probability of a woman belonging to the home birth category. We will use a top-down stepwise method (Nakache and Confais,

2003) to progressively eliminate non-significant variables among the explanatory factors for home births.

The risk difference is calculated from the odds ratio (OR). An OR of less than 1 means that women with the characteristic of the explanatory variable modality under consideration have (1-OR)*100% less risk (or chance) of experiencing the event than women with the reference modality. On the other hand, an OR greater than 1 means that women belonging to the considered modality of the explanatory variable are OR times more likely to experience the event than their counterparts in the reference modality.

The significance of the model parameters is assessed on the basis of the associated critical probability. The model will be considered significant if this probability is less than the significance threshold set at 5%.

<u>**RESULTS AND DISCUSSION**</u>

PART ONE: Spatial and temporal distribution of home births in the Niakhar demographic and health monitoring system.

Chapter 1: Spatial and temporal distribution of home births from 1983 - 2017.

The table above summarizes the number of births from 1984 to 2020, as well as the number of women aged 15 to 40 over the period 1984 to 2016. Analysis of these results shows a clear increase in both the number of births and the population of women aged 15 to 49, testifying to the area's ongoing demographic dynamism. However, there has been a clear decline in the trend of home births over the years, with a notable variation due to disparities in accessibility to maternal health care.

Figure 3Population dynamics in the Niakhar population observatory 1984-2016. (Indices, 1984 = 100)

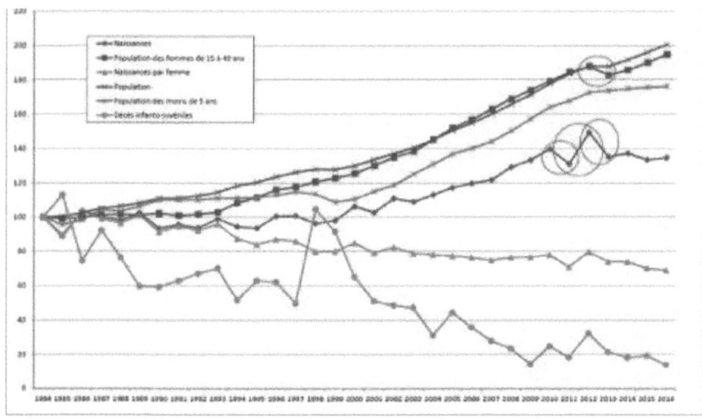

Data sources: Diallo, Guénard, Ojeda Trujillo and Robillard, (2019).

Table 3Number of births and women aged 15-49 for each period (1983-2017).

Year	Number of births	Female population aged 15-49	Average rate of home births per 5-year interval
1983 - 1987	5493	31302	90 %
1988 - 1992	5952	39887	89 %
1993 - 1997	6010	41900	85 %
1998 - 2002	6330	47013	83 %
2003 - 2007	7133	53997	74 %
2008 - 2012	8383	65318	65 %
2013- 2017	8382	70166	49 %
2017	1735	--	43 %
2018	1730	--	33 %
2019	1760	--	26 %
2020	1625	--	20 %

Data sources : Niakhar database and our own calculations

I.1.1. The decade (1983 - 1987) (1988 - 1992).

In the SSDS - de Niakhar, the decade (1983 - 1992) was marked by a very high birth rate and maternal and neonatal mortality, but above all by deliveries carried out in very poor safety conditions. At the beginning of the decade, and particularly during the first five years, the rate of home births was very high in all SSDS - Niakhar villages. In fact, during this period, the average rate of home births was 90% throughout the observatory, with a minimum of 73% and a maximum of 97%. This situation did not really improve during the second period; nevertheless, a slight drop in the rate was noted, with an average of 89% and a minimum of 70%. On the other hand, the maximum rate rose to 100%. Overall, the situation for the decade (1983 - 1992) shows very high proportions of home births in all SSDS - Niakhar villages. Mapping analysis (see figure 2) shows that the lowest rates of home births recorded in the Niakhar observatory during the first period of the decade (1983 - 1992) are located in the villages of Ngayokhème (79%) and Toucar (73%), explained by the fact that two of the observatory's three health posts are established there. However, the highest rates of home births were observed in the villages of Logdir (96%) and Ngangarlame (70%) during this period. Closer examination on a finer scale, that of neighborhoods, does not show great heterogeneity over the whole period. For example, in Ngangarlame, the rates are arranged as follows according to the different neighborhoods: center (pind tok) 96%, Douléme 95%, Pind alang 100%... The rate at village level does not show any discontinuity from the overall rate in the Niakhar SSDS during this period, and rate differences between neighborhoods at village level are not pronounced.

Table 4Average rate, minimum rate, maximum rate for home births (1983/1987-1988/1992).

Year of observation	Number of villages	T. medium	T. maximum	T. minimal
1983 - 1987	30	90 %	73 %	96 %
1988 - 1992	30	89 %	69 %	100 %
Year of observation	Number of hamlets	T. medium	T. maximum	T. minimal
1983 - 1987	166	89 %	25 %	100 %
1988 - 1992	166	87 %	42 %	100 %

Data sources : Niakhar database and our own calculations.

Figure 4Proportions of home births in SSD-Niakhar villages (1983-1987).

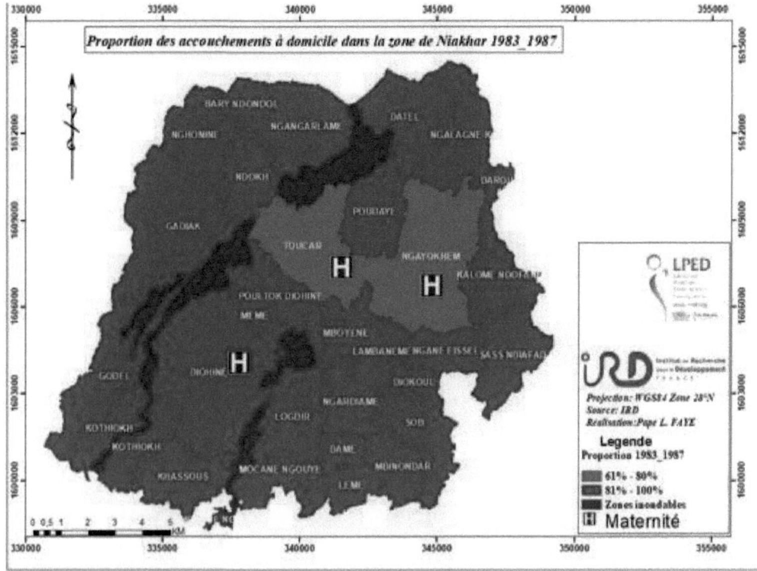

<u>*Data sources*</u> : Niakhar database and our own calculations

Figure 5Proportions of home births in SSD-Niakhar villages (1988-1992)

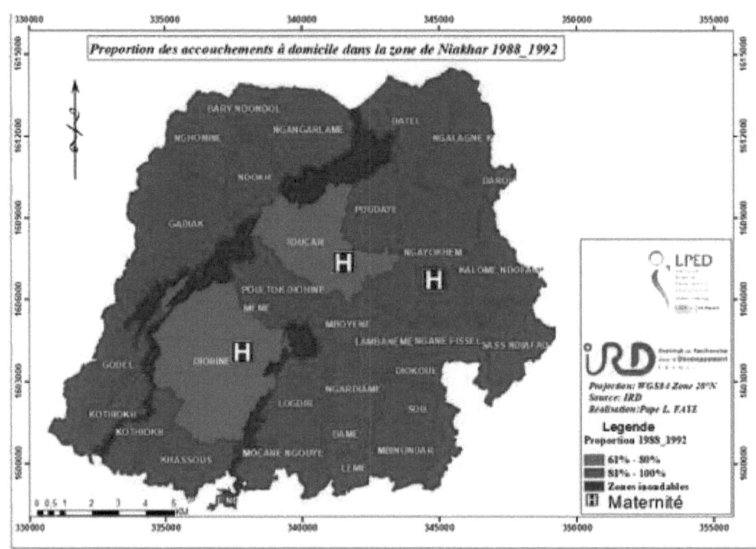

<u>*Data sources*</u> : Niakhar database and our own calculations

I.1.2. The decade (1993 - 1997) (1998 - 2002)

The period (1993 - 2002) shows that assistance by qualified health personnel during childbirth remains very low, especially in villages with no health post. On the other hand, villages close to those with health posts, such as Poudaye, Kalome Ndofane, Ngane Fissel and Darou, have seen a slight upturn in the use of maternity care. These villages are close to the villages of Ngayokhème and Toucar, which are equipped with health posts (see figure 3). Moreover, the rate fell during the first period (1993 - 1997), as well as during the second (1998 - 2002), overall. Between 1993 and 1997, the average home birth rate for the entire observatory was 85%, with a minimum of 57% and a maximum of 97%. For the second period of the decade (1998 - 2002), the average was 83%, with a minimum of 53% and a maximum of 94% (see Table 5). However, it should be noted that the evolution of skilled attendance during childbirth in the Niakhar observatory area shows some irregularity during this period, with oscillations observed in the different villages of the observatory.

Table 5Average rate, minimum rate, maximum rate of home births from (1993 - 1997) - (1998 - 2002)

Year of observation	Number of villages	T. medium	T. minimum	T. maximum
1993 - 1997	30	85 %	57 %	97 %
1998 - 2002	30	83 %	53 %	94 %
Year of observation	Number of hamlets	T. medium	T. minimum	T. maximum
1993 - 1997	166	84 %	25 %	100 %
1998 - 2002	166	82 %	0 %	100 %

Data sources : Niakhar database and our own calculations

Further analysis reveals that the highest rate of home births is recorded in the village of Lambanéme, at 97% during the first period of the decade. In contrast, this rate is observed in the second half of the decade for the village of Bary Ndondol. The villages of Darou, Khassous, Mokane

Gouye, Ngalagne Kop, Ngane Fissel and Sob showed a highly irregular pattern, with no consistency in their evolution. For example, the village of Darou had a home delivery rate of 57% between (1993 - 1997), then increased to 80% during the period (1998 - 2002), as did the village of Ngalagne Kop, which rose from 78% in the first period to 85% in the second half of the decade. In the Niakhar zone, particularly in the villages within the polarization area of the Ngayokhème health post, this irregularity is explained by the limited capacity of this health post, despite its responsibility for a sizeable population (the population of responsibility for the Ngayokhème and Toucar health posts was 29,268 in 2016). In addition, this health post was in an advanced state of disrepair and suffered from the absence of a nurse, which considerably reduced its operation.

Across the SSDS - Niakhar neighborhoods, the rate of home births is highly heterogeneous. During this period, some neighborhoods had a zero rate of home births, while others exceeded 90%. This disparity can be seen in many villages, such as Toucar, where the Kama - Kama district has a rate of 29%, compared with 88% in the Niolélém Niénen district. In Diohine, the Sassar district recorded the lowest rate, at 33%, compared with 77% for the Thitar district. Similarly, in the village of Ngayokhème, this disparity is notable between the Ngayokhème Centre district, which recorded no home births, and the Diayénne district, which achieved a rate of 89%.

Figure 6Proportions of home births at the Niakhar SSDS (1993-1997)

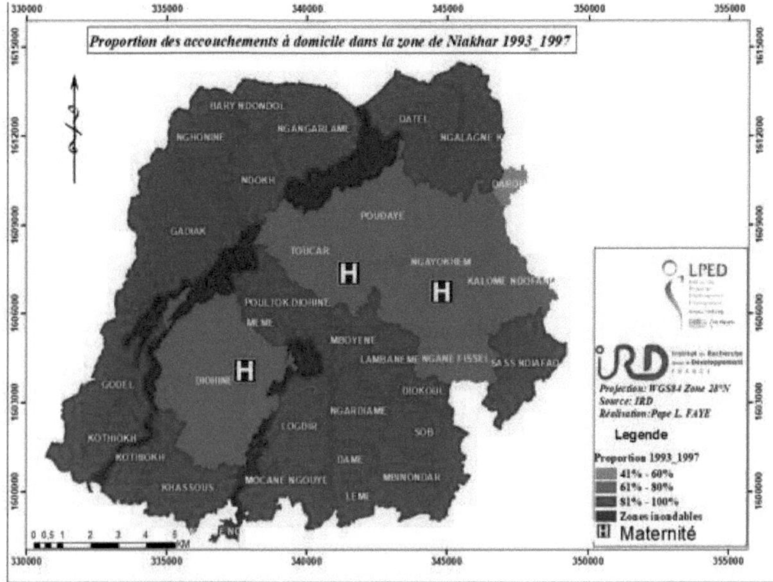

Data sources : Niakhar database and our own calculations

Figure 7Proportion of Niakhar SSDS home births (1998-2002)

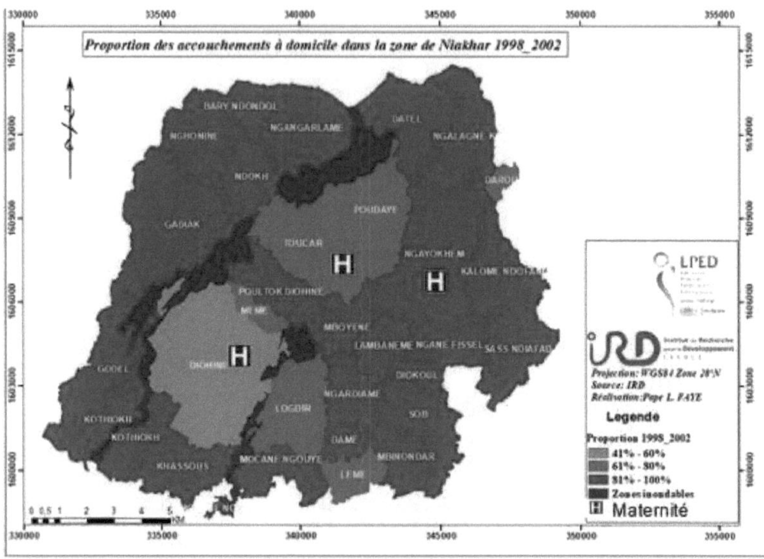

Data sources : Niakhar database and our own calculations.

I.1.3. The decade (2003 - 2007) (2008 - 2012)

In the early 2000s, the trend in home births in the Niakhar SSDS declined considerably. From the start of our observation (1983 - 1987) to the early 2000s, the rate of home births fell by 7% throughout the Niakhar observatory. Analysis of the decade (2003 - 2012) shows that between (2003 - 2007), only ten (10) villages: Gadiack, Godel, Kothiokh, Khassous, Dame, Lambaneme, Diokoul, Poudaye, Datel, Ngalagne Kop remained in the 81% to 100% range, while Diohine and Darou recorded the lowest rates. During this period, ten (10) villages were between 81% and 100%, eighteen (18) villages between 61% and 80%, and two (2) villages between 41% and 60%. In contrast, from 2008 to 2012, only four (4) villages - Godel, Mboyène, Lambanéme and Datel - remained in the

81% to 100% range, fifteen (15) villages between 61% and 80%, and ten (10) villages between 41% and 60%. This drop could be explained by the improvement in healthcare provision in the Niakhar area over this period. Not only through the construction of new health posts, but also through the renovation of two (2) health posts: that of Ngayokhème in 2008, as part of the Belgian Technical Cooperation's (BTC) Projet d'Appui aux Systèmes de Santé des Régions Médicales de Kaolack et Fatick (ASSRMKF), and the private Catholic health post of Diohine, renovated between 2009 and 2010 by the Club International Féminin. These renovations, together with the construction of new consulting rooms and maternity wards, have considerably strengthened the area's healthcare system in response to unprecedented demographic dynamism and growing demand for care.

Table 6Average rate, minimum rate, maximum rate of home births from (2003 - 2007) (2008 - 2012)

Year of observation	Number of villages	T. medium	T. minimum	T. maximum
2003 - 2007	30	74 %	49 %	90 %
2008 - 2012	30	65 %	32 %	83 %
Year of observation	**Number of Hamlets**	**T. medium**	**T. minimum**	**T. maximum**
2003 - 2007	166	76 %	29 %	100 %

2008 - 2012	166	50 %	0 %	100 %

Data sources : Niakhar database and our own calculations

At neighbourhood level, very marked disparities were noted during this period, particularly in the second half of the decade from 2008 to 2012. During this period, the rate of home births ranged from 0% to 100%, depending on the neighborhood. The villages of Mbinondar, Sass Diaffadji and Ngalagne Kop, with their respective neighborhoods of Khoudombedj, Mbin bouré and Ndialo Dialo, had the highest rate, in excess of 90%. Meanwhile, some neighborhoods, such as Ngayokhème, Diobéne de Kalome Ndoffane and others, had a rate of less than 30%. In essence, in the Niakhar observatory, only 38 out of 166 neighborhoods were below the 50% mark for home births during this period.

Figure 8Proportions of home births in the Niakhar SSDS (2003-2007).

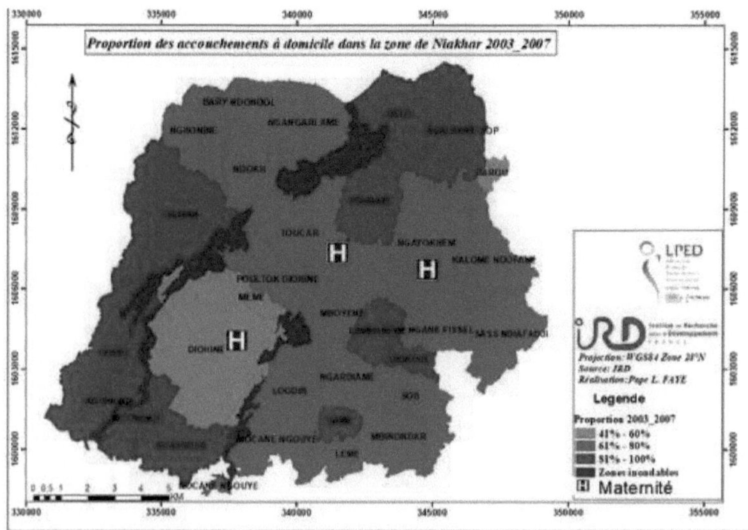

Data sources : Niakhar database and our own calculations

Figure 9Proportions of home births in Niakhar SSDS- (2008-2012).

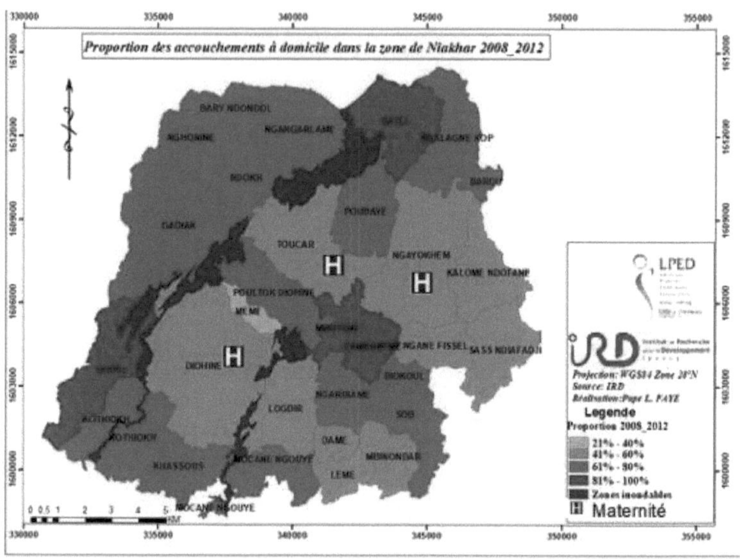

Data sources : Niakhar database and our own calculations

I.1.4. The (2013 - 2017) situation

The period 2013 - 2017 is the last five (5) year interval considered in the analysis of the dynamics of home births in SSDS - Niakhar. This period is notable for its sharp decline in the rate of home births. Between 2008 - 2012 and 2013 - 2017, this rate dropped significantly, recording a 16% decline. At the same time, this period was marked by a strengthening of the healthcare offer with the opening of the Diohine village public health post in 2014, as well as the new Toucar health post to replace the old, dilapidated one established in 1953. Indeed, Toucar has benefited from a new, modern health post with a maternity ward, thanks to the initiative of the village's nationals living in France, as part of the Programme d'Appui aux Initiatives de Solidarité pour le Développement (PAISD).

This period was marked by the disappearance of home birth rates ranging from 81% to 100% throughout the Niakhar area. Only eight (8) villages out of thirty (30), namely Godel, Khassous, Gadiack, Ngonine, Ngangarlame, Datel, Poudaye and Mboyène, remained in the 61% - 80% range. In contrast, thirteen (13) villages, including Kothiokh, Mokane Gouye, Mbinondar, Dame, Ngardiam, Sob, Diokoul, Gane fissel, Poultock, Toucar, Ndokh, Bari Ndondol, Ngalagne kop, showed rates between 41% and 60% (See figure 5). By contrast, the situation at neighborhood level showed disparities in relation to the overall rate for the period. In the Niakhar zone as a whole, 87 neighborhoods recorded rates of between 50% and 100%, while no village had a rate above 80%.

Figure 10Proportions of home births in the Niakhar SSDS- of (2013-2017).

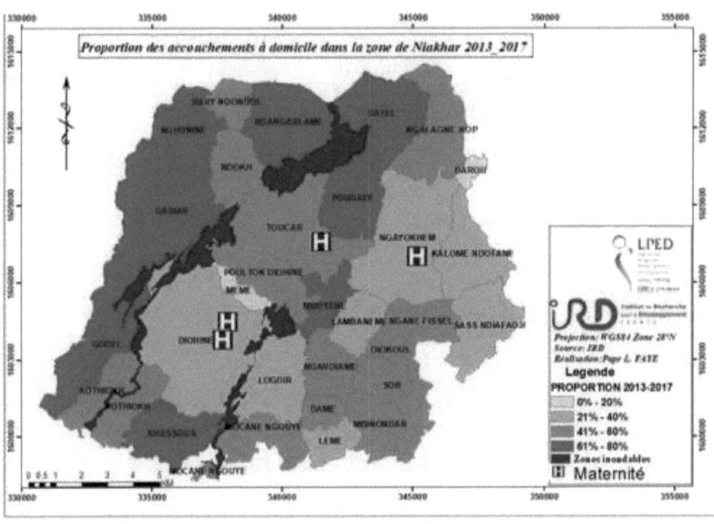

Data sources : Niakhar database and our own calculations.

Table 7Average rate, minimum rate, maximum rate of home births from (2013 - 2017).

Year of observation	T. medium	T. minimum	T. maximum
2013 - 2017	49 %	12 %	78 %
Hamlet	T. medium	T. minimum	T. maximum
2013 - 2017	64 %	0 %	100 %

Data sources : Niakhar database and our own calculations

Chapter 2: Spatial and temporal distribution of home births from 2017 - 2020.

I. The situation at (2017)

In 2017, home births decreased significantly in SSDS - Niakhar. From the start of our observation in 1983 to 2017, the average rate of home births fell by 47%. During this year (2017), it became apparent that more than half of the villages making up the Niakhar observatory had a home birth rate below 50%, with the exception of eight (8) villages: Gadiack, Khassous, Datel, Poudaye, Ndokh, Ngonine, Mboyéne, Kothiokh.

Figure 11Proportions of home births in Niakhar SSDS- in 2017.

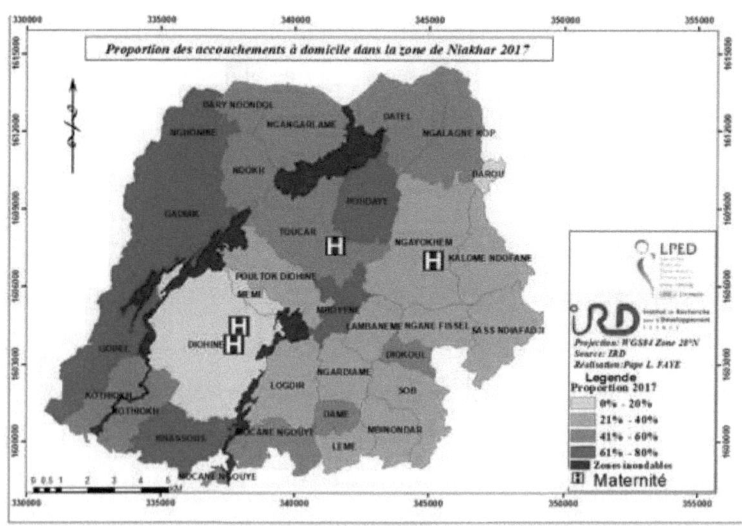

Data sources : Niakhar database and our own calculations

Table 8Average rate, minimum rate, maximum rate for home births in (2017).

Year of observation	T. medium	T. minimum	T. maximum
2017	43 %	0 %	73 %
Hamlet	**T. medium**	**T. minimum**	**T. maximum**
2017	45 %	0 %	100 %

Data sources : Niakhar database and our own calculations

In contrast to the situation in the villages, the neighborhoods of the Niakhar observatory are highly disparate, with 21 out of 166 neighborhoods recording home birth rates of between 81% and 100%. This disparity is all the more notable in that it can be observed even within village neighborhoods with health posts, as in Ngayokhème, where the Centre neighborhood records a rate of less than 10%, while the Mbind Jaga, Monème and Dialo neighborhoods have rates of over 50%.

I.1.6. The situation of (2018)

In 2018, the average proportion of home births in the Niakhar observatory area was 33%, with a maximum of 67% and a minimum of 0%. During this period, some villages, such as Leme, Kalome Ndoffane, Meme and Darou, recorded no home births. However, of the 30 villages in the Niakhar zone, only 9 (Datel, Ngangarlame, Godel, Ngardiam, Ngalagne kop, Ngonine, Sass Ndiafadj, Mboyéne, Kothiokh) recorded a home birth rate of over 50%. In particular, the highest rate was observed in the village of Kothiokh. It should also be noted that these villages, located within a 5 km radius, are partially or totally covered by the influence of the Diohine, Toucar and Ngayokhème health posts.

Table 9Average rate, minimum rate, maximum rate for home births in (2018).

Year of observation	T. medium	T. minimum	T. maximum
2018	33 %	0 %	67 %

Data sources : Niakhar database and author's calculations.

Figure 12Proportion of home births in the Niakhar SSDS in 2018

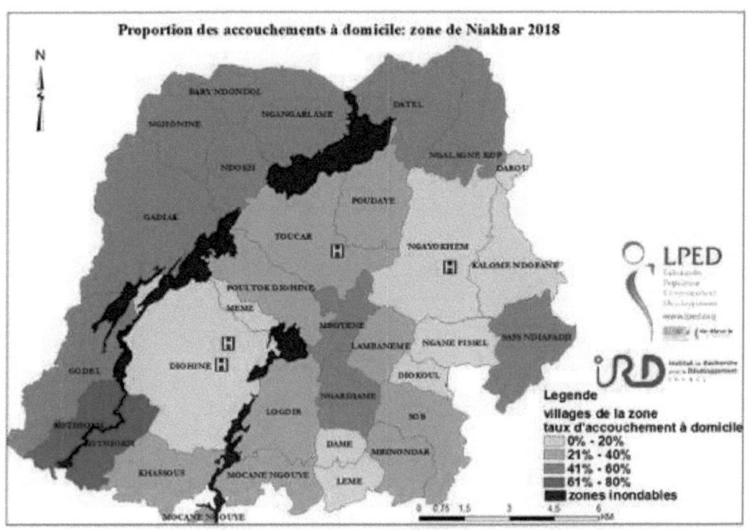

Data sources : Niakhar database and author's calculations.

I.1.7. The situation at (2019)

For 2019, the average proportion of home births in the Niakhar observatory area is 26%, with a maximum of 61% and a minimum of 0%. However, only the village of Darou recorded no home births during this year. Between 2018 and 2019, the average rate of unassisted home births fell by 7%. Nevertheless, the villages of Dame, Ngonine and Godel recorded the highest rates during this year, reaching 50%, 56% and 61% respectively. It is also important to note that the villages of Godel and Ngonine were already in this same category the previous year (2018).

Table 10Average rate, minimum rate, maximum rate for home births in (2019).

Year of observation	T. medium	T. minimum	T. maximum
2019	26 %	0 %	61 %

Data sources : Niakhar database and author's calculations.

Figure 13Proportion of home births in Niakhar SSDS in 2019

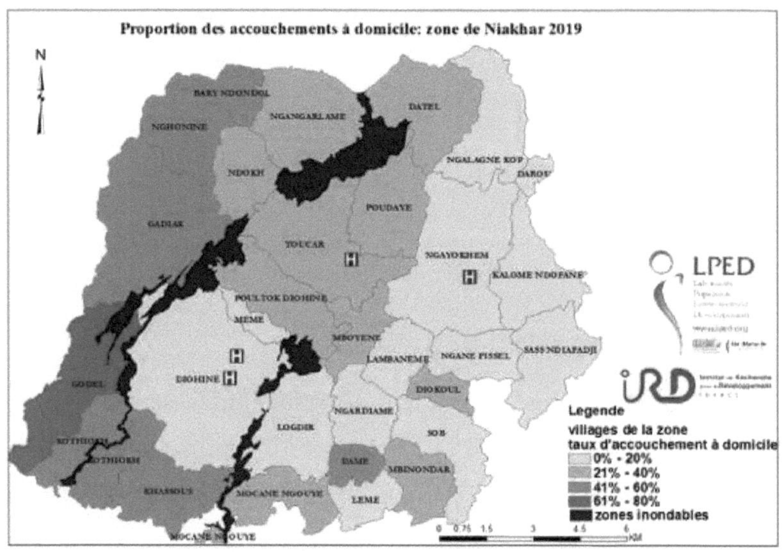

Data sources : Niakhar database and author's calculations.

I.1.8. The situation in (2020)

In 2020, on average, out of 100 births registered by SSDS - de Niakhar, only 20 were not assisted by qualified health personnel, with a maximum of 53% and a minimum of 0%. From 2017 to 2020, the year in which our isolated analysis of years began, the average number of unassisted births fell by 23%. In addition, the villages of Leme, Dame, Diokoul, Meme and Darou recorded no cases of home births during this year. However, only two villages achieved a home birth rate of 50%: Datel and Kothiokh, with 51% and 53% respectively.

Table 11Average rate, minimum rate, maximum rate of home births in (2020).

Year of observation	T. medium	T. minimum	T. maximum
2020	20 %	0 %	53 %

Data sources : Niakhar database and author's calculations.

Figure 14Proportion of home births in the Niakhar SSDS in 2020.

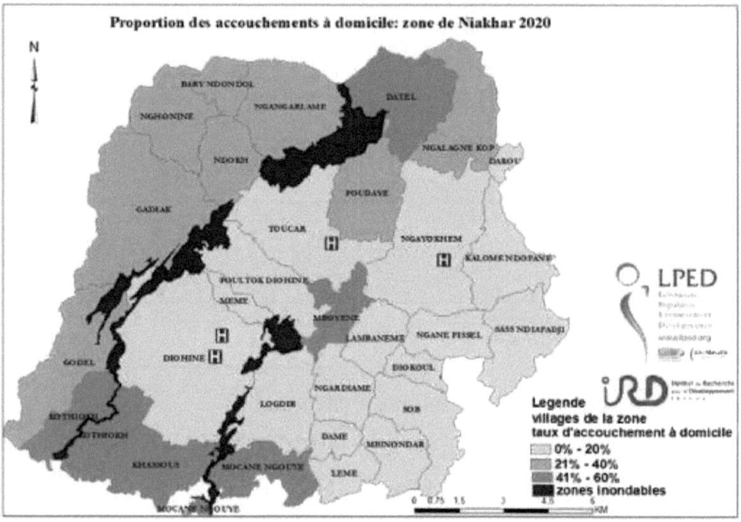

Data sources : Niakhar database and author's calculations.

Table 12Rate of home births in SSD - Niakhar from 1983 to 2017

VILLAGE_NAME	Home births 1983/1987 as % of total	Home births 1988/1992 as % of total	Home births 1993/1997 in % of births	Home births 1998/2002 in	Home births 2003/2007 in	Home births 2008/2012 in	Home births 2013/2017 in	Home births 2017 in	Home births 2018 in	Home delivery 2019 in % of total	Home birth 2020 as % of total
BARY NDONDOL	93,79	90,79	95,68	93,98	73,49	75,78	53,60	45,31	45	41	38
DAME	92,11	87,50	90,24	82,93	80,85	55,56	56,00	50	20	50	0
DAROU	87,50	100,00	57,14	80,00	58,33	68,18	11,76	-	-	-	-
DATEL	96,52	97,08	94,27	91,38	90,36	83,61	70,42	58	50	21	51
DIOHINE	87,89	76,86	66,73	52,80	49,10	44,09	30,24	15,12	10	10	9
DIOKOUL	89,04	89,33	91,94	81,36	80,60	70,83	55,38	50	18	23	0
GADIAK	90,81	89,08	88,65	81,58	82,60	79,77	73,52	73,43	46	46	22
GODEL	92,82	90,56	89,64	85,60	81,89	82,52	66,56	66	51	61	36
KALOME NDOFANE	81,08	82,35	77,99	84,71	70,19	52,41	28,75	25	-	3	8
KHASSOUS	92,50	86,84	87,01	84,31	81,71	75,35	62,43	63,15	31	46	42
KOTHIOKH	92,78	94,78	88,33	85,46	81,21	73,15	58,53	54,16	67	49	53

LAMBANE ME	91,18	80,77	97,32	90,52	83,06	81,90	38,98	24,24	33	8	4
LEME	94,44	91,18	88,00	68,57	73,33	53,49	39,29	40	-	14	-
LOGDIR	95,59	91,39	90,27	76,64	74,39	50,84	34,28	34	33	20	16
MBINOND AR	90,20	89,52	88,79	86,27	62,20	57,45	47,79	33,33	37	24	4
MBOYENE	93,46	89,89	92,41	90,22	79,38	82,09	77,92	73,33	59	29	47
MEME	88,46	96,97	85,19	80,00	52,38	32,14	18,00	11,11	-	11	0
MOCANE NGOUYE	93,69	90,41	84,50	90,00	75,57	63,37	59,20	50	29	31	45
NDOKH	92,05	87,59	95,52	86,21	70,92	75,98	52,22	58,33	44	33	36
NGALAGN E KOP	88,73	94,67	84,97	90,31	86,41	69,96	49,65	52	53	20	22
NGANE FISSEL	87,59	83,33	79,03	88,68	65,00	53,70	40,67	30	17	12	6
NGANGAR LAME	96,12	94,91	90,84	90,60	76,26	78,85	63,66	56,16	50	36	27
NGARDIA ME	92,31	89,81	85,05	82,52	76,03	70,69	54,05	33,33	51	10	14

NGAYOKH EME	79,39	83,75	73,43	87,47	70,50	43,63	27,62	26,37	16	13	8
NGHONINE	92,72	90,20	90,71	88,83	76,69	76,92	63,02	65,09	54	56	39
POUDAYE	84,85	81,52	79,90	78,24	82,53	67,13	65,85	61,22	36	33	24
POULTOK DIOHINE	95,28	89,52	91,34	80,51	79,43	64,20	45,56	31,81	28	23	20
SASS NDIAFADJI	93,08	90,85	86,62	81,56	65,52	50,00	32,08	30,23	57	13	8
SOB	90,77	92,51	84,98	90,15	68,78	61,74	47,84	37,03	30	11	16
TOUCAR	73,36	69,88	64,62	67,88	79,86	54,55	41,14	40,13	24	23	10

Partial conclusion

Spatio-temporal analysis of the dynamics of home birth in the Niakhar population and health observatory reveals a significant improvement in the use of maternal health care. Indeed, at the start of our analysis in 1983, the use of health institutions during pregnancy and childbirth was less than 10% in the Niakhar observatory as a whole. Since then, this situation has progressed very irregularly, varying according to geographical units such as villages and neighbourhoods. This oscillating trend is mainly due to an uneven improvement in healthcare provision, affecting equipment, personnel and quality of care. However, it is important to note that despite a satisfactory maternity care offer in some villages and neighborhoods, the percentage of home births remains high. This is due to a combination of socio-cultural, socio-economic and demographic factors. Thus, it is crucial to adopt a multidisciplinary approach, including field surveys using qualitative data collection and analysis methods, to understand the determinants that influence healthcare utilization. This is the objective of the next stage of our analysis.

**PART TWO: Determinants of home birth in the Niakhar
Demographic and Health Monitoring System**

Chapter 1: Bivariate descriptive analysis

In this section devoted to the second objective of our research, we begin with bivariate descriptive analysis. This method will enable us to assess the correlation between the choice of whether or not to use health facilities during childbirth and the explanatory variables, using the chi-square test (chi2). We will then use the logistic regression method to estimate the associated probabilities.

II.2 Variation according to the woman's economic situation

The woman's economic situation is a factor influencing the use of medical assistance in childbirth. This difference is statistically significant at the 1% level (***=P<0.001). Figure 15 shows a downward trend in the proportion of women not using medical assistance during childbirth as their economic situation improves. Indeed, the proportion of women not assisted at childbirth by qualified health personnel falls from 24% in well-off households to 23% in middle-income households. The highest rate is recorded among women living in poor households, where 34% received no medical assistance during childbirth.

Figure 15Proportion (%) of home births according to the woman's economic situation.

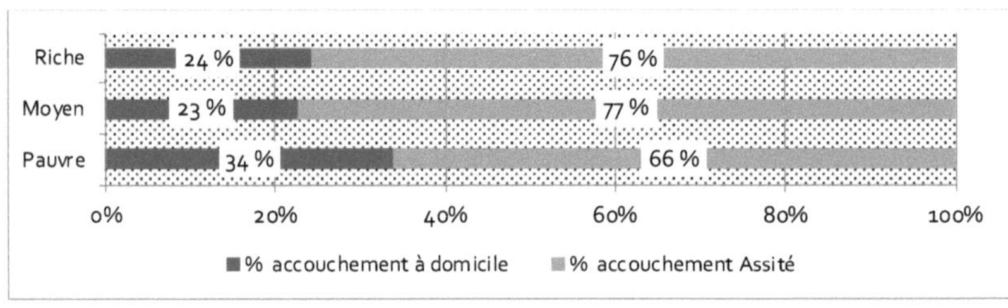

Data sources : Niakhar database and author's calculations.

"It's my husband who supports me for all my health needs, but often I'm forced to manage my health care needs on my own due to lack of means in my household, and that's the main reason I've had many home births." (Ngony Sène, 35, multiparous, uneducated, 5 home births)

"For me, this is one of the main reasons why I always give birth at home: my husband doesn't have the means to provide for my health care during pregnancy and childbirth. Rather than be in the village borrowing money, I prefer to rely on GOD and give birth at home." (Sélbé Ndour, 42, uneducated, multiparous, 7 home births).

II.3. Variation according to the age of the woman.

Medical assistance during childbirth is associated with the woman's age at a significance level of 1% (***, P<0.001). There is a statistical correlation between a woman's age and the absence of assistance from qualified health personnel during childbirth. Indeed, older women are less likely to benefit from assisted childbirth. In the figure above, for women in the [35-49] age bracket, 37% of births took place at home, compared with 31% for those in the [20-34] bracket. However, among women in the [<=20] age bracket, 18% of births took place without the assistance of skilled health personnel.

Figure 16Proportion (%) of home births by woman's age.

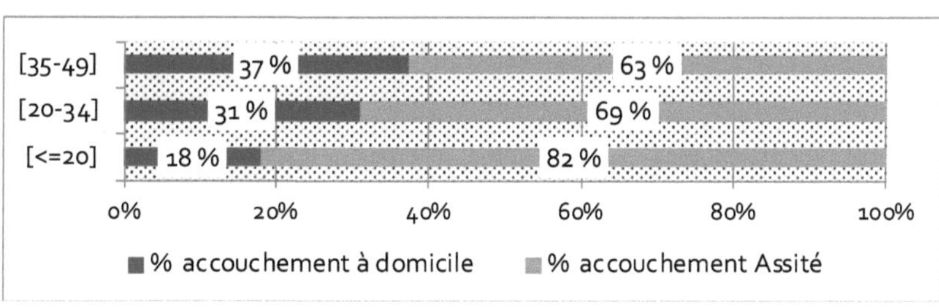

> "Younger mothers are making greater use of maternal health care during childbirth. Most of them are now educated and have a better knowledge of the benefits of assisted childbirth than older mothers, who are sometimes uneducated and know less. What's more, these older women may have a complex about going to the health post because of their age. For example, if an older woman becomes pregnant at the same time as women her children's age, she may avoid prenatal visits and childbirth at the risk of running into them, or even being cared for by younger midwives. In a tight environment, women attach great importance to their image and privacy" (Tabaski Ndour, 47, multiparous, uneducated, 5 home births).

II.4. Variation according to parity achieved.

Achieved parity is associated with non-use of medical assistance in childbirth at the 1% threshold (***=P<0.001). Figure 17 below shows that the proportion of women not receiving medical assistance at childbirth increases with the number of children. Indeed, according to the figure above, women who have had 4 or more children are the least likely to benefit from medical assistance, with 36% of births not attended by qualified personnel. They are followed by those who have had between 2 and 3 children, with a home birth rate of 31%, compared with 18% for women giving birth for the first time.

Figure 17Proportion (%) of home births by parity achieved.

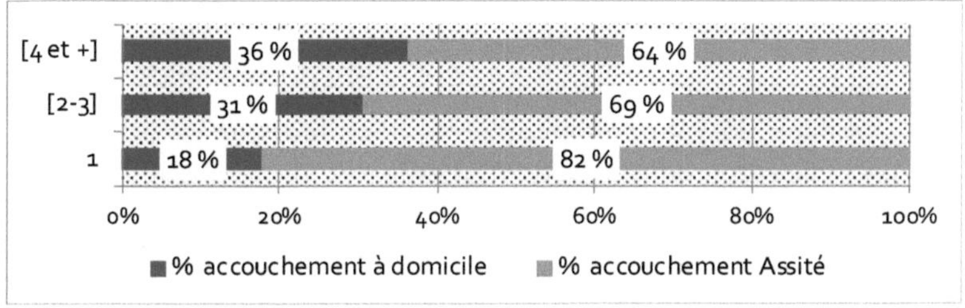

Data sources Niakhar database and author's calculations.

II.5. Variation according to the woman's level of education.

Non-use of medical assistance during childbirth is significantly associated with the woman's level of education at the 1% threshold (***=P<0.001). Figure 18 shows a downward trend in the number of non-medically assisted deliveries as women's level of education increases. However, a slight increase in the rate of home births was observed among university-educated women, compared with those with an intermediate secondary education. On the other hand, women with no form of education are more likely to opt for home births, followed by those with Koranic instruction. Thus, the proportion of uneducated women who give birth without medical assistance is 34%, while it is 31%, 27%, 14% and 18% respectively for women with Koranic, primary, secondary and university education.

Figure 18Proportion (%) of home births according to the woman's level of education.

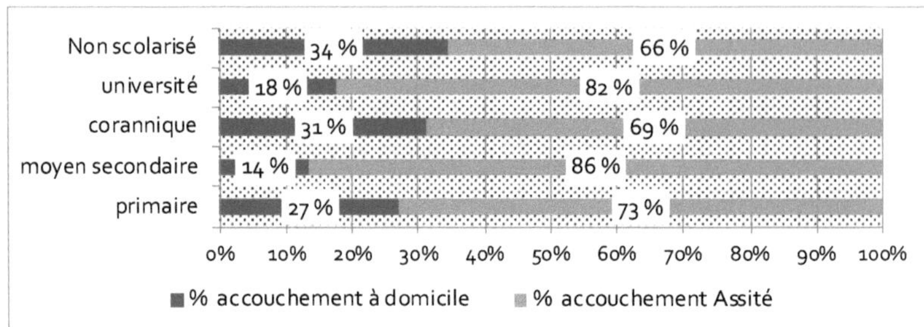

Data sources : Niakhar database and author's calculations.

"I prefer to give birth at the health post because there, the nurses and midwives look after our well-being. For example, at the health post, we are prescribed many medicines that protect both mother and newborn. At home, on the other hand, we have no assistance other than the advice of elderly women using traditional plants and roots with medicinal virtues, the effectiveness of which is sometimes uncertain."

II.6. Variation by marital status

If we look at women's marital status, its relationship with non-use of medical assistance in childbirth is specifically associated with home births at the 1% threshold (***=P<0.001). In fact, home birth is linked to women's marital status. However, the probability of this association being due to chance is 1 in 100. Home birth depends on a woman's marital status, whether she is single, married or divorced. According to our results, married women are more likely to give birth at home, followed by single women, then divorced women. The home birth rate is 33% for married women, 20% for single women and 17% for divorced women.

Figure 19Proportion (%) of home births by marital status.

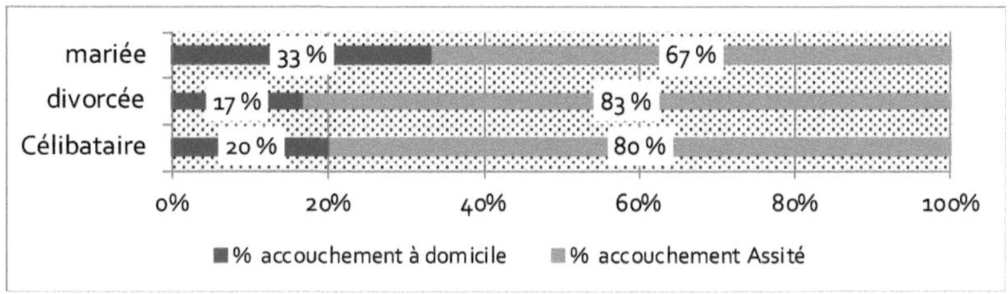

Data source: Field survey and our own calculations

"Single women often give birth at the health post because they don't want to take any risks. Unassisted childbirth is very complicated for a pregnant woman who doesn't live with her husband. In the Serer tradition, housewives who become pregnant can easily give birth at home, as long as they were working during the pregnancy. This enables them to maintain a sufficient level of physical fitness to give birth quietly at home, unlike single women who don't really have any household chores to do." Dibor Ndour, Muslim, 50, uneducated, multiparous, traditional birth attendant

II.7. Variation by number of prenatal consultations

The number of antenatal visits (ANC) is significantly related to non-attendance at birth by a healthcare professional at the 1% threshold (***=P<0.001). In fact, home birth seems to be partially dependent on the number of antenatal consultations carried out. There is a statistical correlation between the number of ANCs and home birth. Among mothers with 2 or fewer ANCs, the rate of home births was 55%, while among those with 3 or more ANCs, the rate dropped to 25%.

Figure 20Proportion (%) of home births by number of ANC.

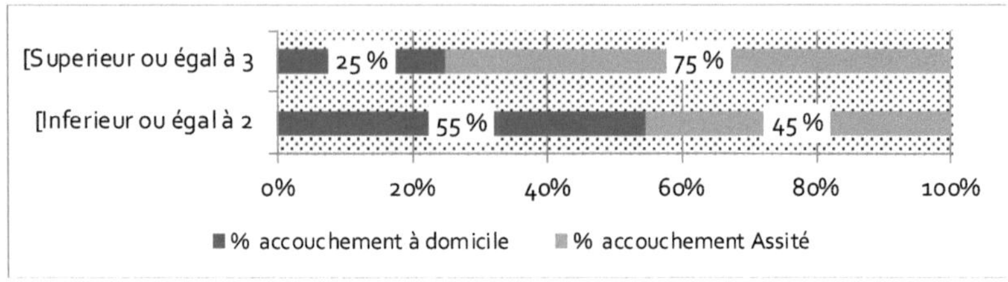

Data sources : Niakhar database and author's calculations.

II.8. Variation according to the woman's caste

The association between the woman's caste and non-use of medical assistance during childbirth is significant at the 1% level (***=P<0.001). In the Niakhar observatory, according to the results in the figure above, 34% of women belonging to the dep DB caste did not use medical assistance during childbirth. On the other hand, the lowest rate of home births is recorded among the non-Sérères, with a home birth rate of 14%. Among the intermediate castes, we find peasant women, griots, unidentified women, artisans and DB thiedo, with proportions of 32%, 31%, 25%, 23% and 18% respectively.

Figure 21Proportion (%) of home births by woman's caste.

Data sources : Niakhar database and author's calculations.

Table 12_:_ Breakdown of workforce by variable.

Variables	Workforce	% home birth
Marital status		
Single	618	20
divorced	6	17
bride	3535	33
Total	4159	31
number of CPN		
[Less than or equal to 2	963	55
[Greater than or equal to 3	3959	25
Total	4922	31
Age group		
[<=20]	590	18
[20-34]	3219	31
[35-49]	1094	37
Total	4903	31
Parity Achieved		
1	689	18
[2-3]	1178	31
[4 and +]	2735	36
Total	4602	32
Economic situation		
Poor	3435	34
Medium	619	23
Rich	868	24
Total	4922	31
Education level		

primary	781	27
secondary means	537	14
Koranic	163	31
university	17	18
Out of school	3415	34
Total	4922	31
Caste		
Farmers	4143	32
DB Tiedo	235	18
dep DB	178	34
Griots	201	31
Artisans	26	23
not tight	35	14
unknown	104	25
Total	4922	31

Data sources : Niakhar database and author's calculations.

Table 13Association of explanatory and dependent variables.

Variables	Parity achieved	Marital status	A woman's age
Pearson Chi2	85,9300***	42,3625***	65,4747***
Variables	The number of CPN	Woman's level of education	Women's economic situation
Pearson Chi2	315,9572***	111,7378***	52,2502***
Variables	the woman's caste.		

Pearson Chi2	27.6397***		

The bivariate analysis is performed considering the following statistical probability levels: Not significant (ns = P > 0.05); significant (** = 0.01 < P < 0.05); highly significant (*** = P < 0.001).

Partial conclusion

The descriptive bivariate analysis shows that all the selected variables (parity achieved, marital status, woman's age, number of prenatal consultations, woman's level of education, woman's economic situation) are statistically linked to the use or non-use of a health facility during childbirth. On the other hand, the influence or link established between the dependent variable and the other explanatory variables is not intended to demonstrate a causal relationship, but rather to visualize or determine a correlation.

Chapter 2: Analysis and interpretation of the results of the binary logistic regression model.

Descriptive bivariate analysis, using the chi-square test, established associations between the dependent and explanatory variables. However, the results do not allow us to establish causal relationships. Thus, the aim of this section is to highlight the factors associated with the phenomenon, taking into account the effects of different explanatory variables. To this end, we will use binary logistic regression. This part is structured into 3 sections: the first provides an overview of the analytical models and the conditions for their validity, the second highlights the factors that explain non-use of medical assistance in childbirth, while the third is devoted to a discussion of the results.

III.1 Presentation of the analysis model

To identify the factors that explain non-use of medical assistance in childbirth, and to understand the mechanisms by which each independent variable influences its occurrence, several models have been set up. We will use the step-by-step top-down method (Nakache and Confais, 2003) to progressively eliminate non-significant variables. We'll start with a model containing only the dependent variable, then gradually introduce the various independent variables.

III.2. Matching models to data

The adequacy of models to the data is assessed using the probability of the chi-squared test. A model is considered adequate if this probability is lower than the chosen threshold. In this study, the significance threshold is set at 5%. The various probabilities associated with the chi-squared tests of our models are below this threshold. This indicates that, on the whole, the variables used explain the phenomenon.

III.3 Quality of fit of data to models

To illustrate the causal structure of the study, we evaluate the predictive and discriminative power of the models. The predictive power of the models is determined using the 'estat classification' procedure in STATA software. This test highlights the proportion of good predictions generated by the model. Thus, using the probability from the regression as a measure, we will be able to predict on a scale of 100 the number of women who do not seek medical assistance during childbirth. When this value exceeds 50%, the model correctly predicts the phenomenon.

Table 14Correct prediction rates for the binary logistic model

Number of correct predictions	Number of observations	Correct prediction rate
2696	3811	70.74%

Data sources : Niakhar database and author's calculations.

Secondly, the discriminating power of the model is evaluated using the ROC (Receiver Operating Characteristics) curve. The area under the ROC curve is used to assess the model's accuracy in terms of discriminating power. A high discriminating power of the model is reflected by a value of the area under the ROC curve tending towards 1. Figure 22 shows the ROC curve of our saturated model. The area under the model's ROC curve is equal to 0.6947. The low discriminatory capacity of the models suggests that there are other explanatory variables for non-use of medical assistance in childbirth that have not been taken into account in these models.

Figure 22ROC curve showing the discriminative power of our model.

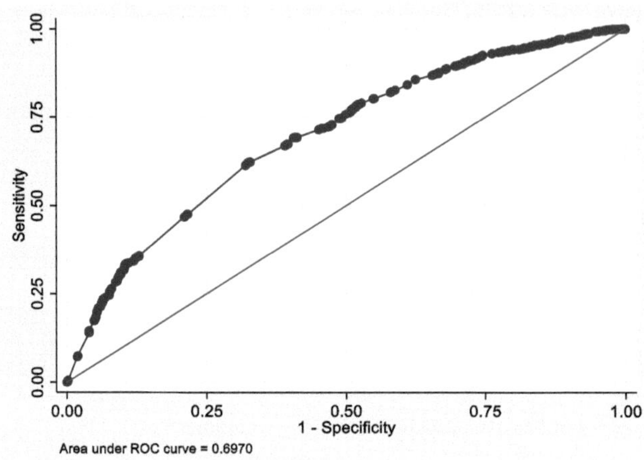

Data sources : Niakhar database and author's calculations.

III.4 Explanatory factors for non-use of medical assistance in childbirth

Table 15Odds ratio and marginal effect estimates for factors explaining non-attendance by healthcare personnel during childbirth.

Place of birth	Gross effect	Net effect		
		Odds Ratio	**p-value**	**Marginal effects**
Marital status				
Ref: Single				
Divorced	0.788 ns	0.339 ns	0,341	-0.157
Bride	1.968***	1.265 ns	0,092	0.044
Education level				
Ref: Primary				

Middle-Secondary	0.428***	0,587***	0,005	-0.092
Koranic	1,241 ns	1.475 ns	0,067	0.079
University	0.584 ns	0.638 ns	0,674	-0.079
Out of school	1.433***	1.211ns	0,059	0.037
Number of CPN : **Ref: <=2**				
>=3	0.278***	0,285***	0,0000	-0.278
Age of the woman **Ref : [<=20]**				
[20-34]	2.022***	1.255 ns	0,154	0.043
[35-49]	2.674***	1.322 ns	0,122	0.053
Parity reached **Ref: 1**				
[2-3]	2.025***	1,638***	0,001	0.090
[4 and +]	2.602***	1,793***	0,0000	0.108
Socio-economic level **Ref: Poor**				
Medium	0.569***	0,563***	0,0000	-0.109
Rich	0.625***	0,551***	0,0000	-0.113
A woman's castle Ref: Paysan				
DB Tiedo	0,468***	0,536***	0,001	-0,112
dep DB	1.094ns	0.896ns	0.585ns	-0,021
Griots	0.982ns	1.042 ns	0.825ns	0,008
Artisans	0.645ns	0.494ns	0.17ns	-0,125
not tight	0,358***	0.304ns	0.061ns	-0,19
unknown	0.717ns	0.733ns	0.253ns	-0,19
Constant		0,569***	0,006	

Prob > chi2	0,0000	
R2 username	0,087	
Comments	3811	

Data sources : Niakhar database and author's calculations.

Logistic regression analysis is performed considering the following statistical probability levels: Not significant (ns=P>0.10); significant at the 5% level (**=0.01<P<0.05); significant at the 10% level (*=0.01<P<0.10).

III.4.1. Interpretation of marginal effects

The result of the above table shows that the woman's level of education, the number of prenatal consultations, the parity achieved, the woman's economic status, and the woman's caste have significant marginal effects on the use of skilled health personnel at the time of birth. On the other hand, the woman's age and marital status have marginally insignificant effects on the use of health facilities during childbirth.

Level of education, number of antenatal visits, economic status and caste have marginal negative effects on the likelihood of a woman giving birth at home.

More precisely, a woman's level of education has a negative effect on the probability of her giving birth at home. All other things being equal, a one-unit increase in a woman's level of education results in a 9.2% drop in the probability of giving birth at home among women educated to intermediate secondary level. This probability is not significant for university-educated, Koranic-educated and uneducated women.

Similarly, there is a 27.8% drop in the probability of giving birth at home if the woman has had >=3 prenatal consultations. We also note that only among women with a stable economic situation does the probability of giving birth at home decrease. For women of average economic status, the probability of giving birth at home fell by 10.9%, and by 11.3% for wealthy women. Similarly, a woman's caste also has a negative influence on home birth. Indeed, a one-unit increase in a woman's caste results in an 11.2% drop for women belonging to the DB Tiedo caste. This probability is not significant for the other caste groups (dep DB, Griots, artisans, non Serrére, not known). On the other hand, parity has a positive marginal effect on the probability of giving birth at home. For women who have had 2 to 3 children, the probability of giving birth at home increases by 9%, and by 10.8% for those who have already given birth 4 times or more.

III.4.2. Interpretation of Odds-Ratios

According to the results of our logit model, the odds ratios are as follows: when variables such as level of education, number of prenatal consultations, woman's economic status or woman's caste increase by one, the probability that a woman will be assisted by skilled health personnel at the time of birth increases.

For example, women educated to medium-secondary level are 0.587 times more likely to be attended by skilled health personnel during birth than women educated to primary level. However, this probability is not significant for university-educated, Koranic-educated or uneducated women. Similarly, women with >=3 prenatal consultations were 0.285 times more likely to give birth in a health facility than those with <=2 prenatal consultations.

In terms of economic status, wealthy and middle-income women are respectively 0.563 and 0.551 times more likely to benefit from medical assistance than poor women. For those belonging to the "DB Tiedo" caste, they are also 0.536 times more likely not to give birth at home than peasant women.

However, a one-unit increase in parity leads to a decrease in the probability of using a qualified health facility during birth. In fact, this decrease is 1.638 times for women who have had 2 to 3 children, and 1.793 times for those who have had 4 or more children.

III.4.3. Prioritization of explanatory factors

To guide actions aimed at reducing the number of home births, it is essential to identify the factors influencing this trend. We now need to determine the explanatory power of each factor, in order to rank them in order of importance. This is done by calculating the absolute and relative contributions of each of the determining variables in explaining non-use of medical assistance in childbirth. The absolute contribution is given by the following formula :

$$C\,variables = \frac{Chideux\ du\ modéle\ saturé - Chideux\ du\ modéle\ sans\ la\ variable}{Chideux\ du\ modéle\ saturé}$$

The relative contribution of a variable is obtained by dividing its absolute contribution by the total sum of absolute contributions. The factors with the greatest contribution have the greatest influence. Only factors that are decisive in the explanation at the 5% threshold are taken into account in the ranking.

Table 16Hierarchy of determining factors.

Variables	Chi-square of saturated model	Chi-square of the saturated model without the variable	Contribution absolute (%)	Contribution (%) contribution (%)	Rank
Number of prenatal consultations	421.06	209.55	50 %	69,14 %	1
Women's economic situation	421.06	370.09	12,10 %	16,73 %	2
Education level	421.06	396.42	5,85 %	8,09 %	3
A woman's castle	421.06	402.35	4,44 %	6,14 %	4
Parity reached	421.06	455.43	-0,081 %	-0,11 %	5

Data sources : Niakhar database and author's calculations.

Table 17Comparison of hypotheses and results.

The assumptions of the study are :	Conclusion
H1: Level of education has a significant influence on non-use of medical assistance in childbirth. We expect uneducated	

women to have a higher risk of not using medical assistance in childbirth than their educated counterparts.	Confirmed
H2 We assume that a woman's age influences non-use of medical assistance in childbirth, translating into a higher risk for older women of not using medical assistance in childbirth compared to those who are younger.	Unconfirmed
H3: The woman's standard of living has an influence on non-use of medical assistance at childbirth, with women from poor households more likely not to use medical assistance at childbirth than those living in wealthy households.	Confirmed
H4 We assume that the parity reached by the woman influences non-use of medical assistance during childbirth; women who have had 4 or more children are more likely not to use medical assistance during childbirth than those giving birth for the first time.	Confirmed
H5: We assume that women from the peasant caste are less likely to use assisted childbirth than their counterparts from other castes.	Confirmed
H6: We assume that marital status affects the use of medical assistance during childbirth, with married women more likely not to use medical assistance during childbirth than unmarried women.	Unconfirmed

H7: We assume that women with less than or equal to 2 antenatal visits have a higher risk of not using medical assistance during childbirth than those with 3 or more antenatal visits.	Confirmed

Study limits

Like all research, this study has important limitations. Firstly, the limitations of this study lie in its first part. The retrospective data on home births, which we exploited from the Niakhar database, even cover the period to the year 2021. However, our study was limited to the year 2020 due to a fairly substantial number of missing data for the latter period. The availability of recent data for 2021 would undoubtedly have enabled us to gain a better understanding of current dynamics.

The second part of this research also identifies a number of limitations, particularly with regard to the explanatory variables. One such limitation is the absence of the 'gender' variable, which reflects women's status in society. According to Rakotondrabe (2004), a woman's status is defined as "a social position providing her with a certain prestige within society, influencing her access to resources, her ability to control these resources and her decision-making power within the household". Similarly, Sidibé (2019) points out that "frequent discussions within the couple on family planning lead to consideration of the woman's concerns in this area as well as her health". Integrating this variable proved complex given the specific social context of our research field and the delicate nature of this variable, as well as other factors such as religion, financial and geographical accessibility, spouse's profession, quality of obstetric care, etc.

Similarly, the variable 'timing of antenatal care request in last pregnancy' could have been an essential element. Indeed, according to Beninguisse et al (2007), home births are strongly related to the time elapsed between the first prenatal consultation and delivery, with a significantly higher frequency among women who had their first consultation beyond the first trimester of pregnancy. Early prenatal consultations are therefore crucial. However, tracing the therapeutic pathways of the women in this study proved difficult due to gaps in the retention of antenatal consultation diaries and possible biases in interview data collection, potentially altering the quality of the information gathered.

GENERAL CONCLUSION

The main objective of this research was to study the spatial and temporal distribution of home births in order to identify at-risk areas in the Niakhar SSDS, and then to determine and analyze the factors contributing to this phenomenon.

Initially, we used retrospective data from the Niakhar SSDS reproductive health database to calculate the proportion of home births in the 30 villages and 166 hamlets of the Niakhar population and health observatory.

Our analysis revealed a very heterogeneous distribution of home births in the Niakhar zone, varying according to the different geographical areas, but following a logic of care provision that was itself heterogeneous. For example, all villages in the Niakhar area with a health post had significantly lower delivery rates than those without. However, a more in-depth analysis of the neighborhoods of these villages with health posts revealed disparities in home birth rates.

Following data collection and processing, it emerged that the determinants of home birth were mainly linked to socio-economic, socio-demographic and socio-cultural characteristics, and to factors affecting the supply of care.

The results of the binary logistic regression model showed that women educated to medium-high school level were 0.587 times more likely to be assisted by skilled health personnel during childbirth than women educated to primary school level. However, this probability was not significant for university-educated, Koranic-educated or uneducated women. Similarly, women with >=3 prenatal consultations were 0.285 times more likely to give birth in a health facility than those with <=2 consultations. Wealthy and middle-income women were respectively 0.563 and 0.551 times more likely to receive medical assistance than poor women. In addition, women belonging to the "DB Tiedo" caste were also 0.536 times more likely not to give birth at home than peasant women. On the other hand, an increase in the parity index led to a decrease in the propensity to use a qualified health facility during childbirth. This decrease was 1.638 times for women with 2 to 3 children, and 1.793 times for those with 4 or more children.

In sum, these results of the binary logistic regression model highlight that the woman's level of education, the number of prenatal consultations, the parity achieved, the woman's economic situation and the woman's caste are the most significant determinants explaining discontinuity in the use of a health facility during childbirth in the Niakhar demographic and health monitoring system.

However, these results relate to a limited number of variables. A more exhaustive analysis including a larger set of relevant variables would allow us to confirm the relationships identified here. This is what we plan to achieve in our future research perspectives."

BIBLIOGRAPHICAL REFERENCES

1. **Ahmed, M.A.A. (2019)** "Les déterminants du recours ou non à l'accouchement assisté par les femmes nomades de Gossi au Mali et les stratégies potentielles pour le faciliter.", p. 242.

2. **Alain, T.N.G. (2015)** "L'ACCOUCHEMENT A DOMICILE AU CAMEROUN", p. 34.

3. **Akoto E. M. (1993)**, Déterminants socioculturels de la mortalité des enfants en Afrique Noire. Hypothèses et recherche d'explication, Louvain-la-Neuve, Académia, 269p.

4. **Ahmed, M.A.A. (2019)** "Les déterminants du recours ou non à l'accouchement assisté par les femmes nomades de Gossi au Mali et les stratégies potentielles pour le faciliter.", p. 242.

5. **Baya, B. (1998).** Instruction des parents et survie de l'enfant au Burkina Faso: Cas de BoboDioulasso. Les dossiers du CEPED, (48), 27.

6. **Beninguisse, G., Nikièma, B. and Fournier, P. (2003)** "L'accessibilité culturelle : une exigence de la qualité des services et soins obstétricaux en Afrique", 19, p. 24.

7. **Burrous, H.A. (2020)** "Influence interpersonnelle et soins de maternité institutionnels: Influence des individus, des quartiers et des réseaux sociaux sur la perception du besoin de soins médicaux à la naissance à Niakhar, au Sénégal", p. 49.

8. **Bénié Bi Vroh, J. *et al.* (2009)** "Prévalence et déterminants des accouchements à domicile dans deux quartiers précaires de la commune de Yopougon (Abidjan), Côte d'Ivoire", *Santé Publique*, 21(5), p. 499.

9. **Ba Gning, S. and Sandberg, J. (2018)** "Chapter 15. Tomber malade et en guérir sans aller au dispensaire", in Delaunay, V., Desclaux, A., and Sokhna, C. (eds) *Niakhar, mémoires et perspectives*. IRD Éditions, pp. 295-309.

10. **Barlet, M. *et al* (2012)** 'L'Accessibilité potentielle localisée (APL): une nouvelle mesure de l'accessibilité aux médecins généralistes libéraux', p. 8.

11. **Beninguisse G. (2003),** Entre tradition et modernité. Fondements sociaux de

la prise en charge de la grossesse et de l'accouchement au Cameroun, Académia Bruylant/L'Harmattan, Louvain-la-Neuve/Paris, 298p.

12. **Beninguisse G.et Bakass F. (2007)** " Santé de la reproduction et statut des femmes dans le ménage : l'exemple du Cameroun et du Maroc " in Genre et société en Afrique, implication pour le développement, paris, les cahiers de l'INED. PP 395-415.

13. **Beninguisse G. et al (2009),** "La discontinuité des soins obstétricaux en Afrique subsaharienne" in Mémoires et Démographie, Regards croisés au Sud et au Nord; les cahiers du ciéq, pp. 366 - 385.

14. **Chippaux, J.-P. (2005)** *Recherche intégrée sur la santé des populations à Niakhar (Sahel sénégalais)*. Paris: IRD Éditions.

15. **Carmen, M. (2006)** "USE OF HEALTH CARE SERVICES IN CAMEROON", p. 12.

16. **Delaunay V., Desclaux, A. and Sokhna, C. (2018)** "Niakhar, mémoires et perspectives: recherches pluridisciplinaires sur le changement en Afrique", p. 536.

17. **Delaunay, V. (2018)** "La situation démographique dans l'Observatoire de Niakhar: 1963-2014", p. 90.

18. **Delaunay, V.** *et al.* **(2019)** "The Niakhar Social Networks and Health Project", *MethodsX*, 6, pp. 1360-1369.

19. **Duchaine, F.** *et al* **(2020)** "La situation démographique et sanitaire dans l'Observatoire de Bandafassi: 1970-2016", p. 109.

20. **Diallo, H. (2018)** "Spécialité: Politiques publiques Mémoire de recherche de Master 2", p. 80.

21. **Diallo B. et al (1999),** "Problèmes médicaux et culturels de l'inadéquation entre le taux de consultation prénatale et d'accouchements assistés dans les quatre régions naturelles de la Guinée", Médecine d'Afrique noire, 46 pp. 32-39.

22. **Dieng, M.** *et al.* **(2014)** "Déterminants de la demande de soins en milieu péri-urbain dans un contexte de subvention à Pikine, Sénégal", p. 28.

23. **Dany, L. (2017)** "Qualitative content analysis of social representations", p. 38.

24. **De Souza A. O. (1995),** La maternité chez les Bijagode Guinée-Bissau : une analyse épidémiologique et son contexte ethnologique, Les Etudes du CEPED n°9, Paris, 114p.

25. **Faye, A. *et al.* (2010)** "Factors determining the place of delivery among women who received at least one prenatal consultation in a health facility (Senegal)", *Revue d'Épidémiologie et de Santé Publique*, 58(5), pp. 323-329.

26. **Faye, S.L. (2008)** "Devenir mère au Sénégal : des expériences de maternité entre inégalités sociales et défaillances des services de santé", *Cahiers de Santé*, 18(3), pp. 175-183.

27. **Fournier P., and HADDAD S., (1995),** "Les facteurs associés à l'utilisation des services de santé dans les pays en développement" in La sociologie des populations, Montreal, PUM AUPELF-UREF, pp. 289-325.

28. **Gastineau, B. and Golaz, V. (2016)** "Être jeune en Afrique rurale Introduction thématique", *Afrique contemporaine*, 259(3), p. 9.

29. **Garenne, M. (2019)** "Morbidity and causes of death: the contribution of the demographer", p. 18.

30. **Garenne, M. *et al.* (2018**) "Chapter 7. Fifty years of mortality transition in Niakhar (1963-2012)", in Delaunay, V., Desclaux, A., and Sokhna, C. (eds) *Niakhar, mémoires et perspectives*. IRD Éditions, pp. 151-170.

31. **Guyavarch, E. (2007)** "En Afrique, des suivis de population sur le terrain pour mieux saisir les tendances démographiques", p. 4.

32. **Houndole, N. and Künzi, S. (2014)** "Étudiante Bachelor - Filière Sage-femme", p. 149.

33. http://dhs program.com

34. http://collections.banq.qc.ca/ark:/52327/3995348

35. **Institut national d'excellence en santé et en services sociaux (Québec), C., Brigitte, Auclair, Yannick, Guise, Michèle de, Institut national d'excellence en santé et en services sociaux (Québec) and Direction des services de santé et de l' évaluation des technologies (2019)** *Sécurité du lieu et conditions de succès de l' accouchement vaginal après une césarienne: avis.*

36. **Jaffre Y. and Prual (1993)**, "Le corps des sages-femmes entre identités professionnelles et sociales", Sciences sociales et santé, vol XI, n°2, pp. 63-80.

37. **Jodelet, Denise (2003)**. Les représentations sociales (7th ed.). Paris: Presses universitaires de France.

38. **Kanté, A.M. and Pison, G. (2010)** "La mortalité maternelle en milieu rural sénégalais. L'expérience du nouvel hôpital de Ninéfescha", *Population*, 65(4), p. 753.

39. **Kwete, M.B. (2016)** "Factors favoring home deliveries in the Rural Health Zone of Lemera in South Kivu, DR Congo [Factors associated with home delivery in Rural Health Zone of Lemera, DR Congo]", 17(4), p. 7.

40. **Kalilou, O. (2019)** "Factors explaining home birth in the village of Namassi (North-East cote d'ivoire)", p. 14.

41. **Leroy, O. and Garenne, M. (1980)** "La mortalité par tétanos néonatal : la situation à Niakhar au Sénégal", p. 9.

42. **Messi, E. and Yaye, W. (2017)** "Contraintes À L'accès Aux Soins De Santé Maternelle Dans La Ville De Maroua", *The International Journal of Engineering and Science,* 06(01), pp. 13-21.

43. **Munyemana, M. and Kakoma, J.B. (2010)** "Facteurs influençant le lieu d'accouchement dans le district de Nyaruguru (Province du sud du RWANDA)", 68, p. 5.

44. **MASUY-STROOBANT (1996),** "Théories et schémas explicatifs de la mortalité des enfants", in: Casseli, Vallin et Wunsch, G. (eds), Démographie : analyse et synthèse. Causes and consequences of demographic change, Dipartimento di Scienze Demografiche and CEPED, Rome and Paris, San Miniato, vol. 2, pp. 193-207.

45. **Mitchell, M.N. (2004)** *A visual guide to stata graphics*. College Station, Tex: Stata Press.

46. **Mizrahi, Andrée and Mizrahi, Arié (2010)** 'La densité répartie : un instrument de mesure des inégalités géographiques d'accès aux soins', *Villes en parallèle*, 44(1), pp. 94-113. doi:10.3406/vilpa.2010.1474.

47. **MEBTOUL, M. (1993),** La santé au quotidien : le dispensaire du quartier d'El

Hamri (ORAN), in: Sciences sociales et santé, Vol. XI, n°2, pp. 41-62.

48. **Nkurunziza, M. (2015)** "Accoucher à domicile malgré la gratuité des soins: Le cas du milieu rural burundais", *Autrepart*, 74-75(2), p. 85.

49. **Ngom, N.F. (2017)** "L'assistance médicale à l'accouchement au Sénégal", p. 373.

50. **Ngoné Déguène Samb and Papa Sakho (2012)** "Déterminants de l'utilisation des services de la santé de la reproduction (SR) par les populations de transhumants pastoraux de la région de Matam".

51. **Nkoumou Ngoa, G.B. (2020)** "GRATUITÉ DES SOINS ET UTILISATION DES SERVICES DE SANTÉ MATERNELLE - UNE ANALYSE D'IMPACT AU *SÉNÉGAL*" *l'actualité économique*, 96(2), p. 159.

52. **Ndiaye, P. *et al.* (2005)** "Déterminants socioculturels du retard de la 1re consultation prénatale dans un district sanitaire au Sénégal", *Santé Publique*, 17(4), p. 531.

53. **Nankwanga, A. (2004),** Factors influencing utilisation of postnatal services in Mulago and Mengo hospitals, Kampala, Uganda, Minithesis for a Master of Science Physiotherapy.

54. **Niang, A. and Handschumacher, P. (1998)** "et recours aux soins de santé primaires", p. 25.

55. **Olivier de Sardan Jean-Pierre, Adamou Moumouni, Aboubacar Souley, 1999,** "L'accouchement c'est la guerre - De quelques problèmes liés à l'accouchement en milieu rural nigérian"], pp. 1-125.

56. **Ochako R., Fotso J.-C., Ikamari L., Khasakhala A. [2011],** "Utilization of maternal health services among young women in Kenya: Insights from the Kenya demographic and health survey, 2003", *BMC pregnancy and childbirth*, vol. 11, no. 1, pp. 1-9: www.biomedcentral. com/1471-2393/11/1 (page consulted February 3, 2011).

57. **Magadi M. A., Agwanda A. O., Obare F. O. [2007],** "A comparative analysis of the use of maternal health services between teenagers and older mothers in sub-Saharan Africa: Evidence from demographic and health surveys (DHS)",

Social science and medicine, no. 64, pp. 1311-1325.

58. **Bénie Bi Vroh Joseph, Timbre Issaka, Zengbe Acray Pétronille, Gueu Doua Judith , N'cho Simplice Dagnan and Tagliante-Saracino Janine, Santé Publique 2009/5 (Vol. 21), 2009** " Prévalence et déterminants des accouchements à domicile dans deux quartiers précaires de la commune de Yopougon (Abidjan), Côte d'Ivoire ", Santé Publique, volume 21, no. 5, p. 499-506.

59. **Pison, G.** *et al* **(1989**) "L'influence des changements sanitaires sur l'évolution de la mortalité : le cas de Mlomp (Sénégal) depuis 50 ans", p. 36.

60. **Pison, G.** *et al* **(2000)** "La mortalité maternelle en milieu rural au Senegal", *Population (French Edition)*, 55(6), p. 1003.

61. **Pison, G.** *et al* **(1989)** "L'influence des changements sanitaires sur l'évolution de la mortalité : le cas de Mlomp (Sénégal) depuis 50 ans", p. 36.

62. **Pebley Anne, Goldman Noreen, Rodriguez German, 1996,** "Prenatal and delivery care and childhood immunization in Guatemala: do family and community matter", Demography, vol. 33, no. 2, pp. 231-247.

63. **Rakotondrabe, F. P. (2001, July).** Contribution du genre à l'explication de la santé des enfants: cas de Madagascar. In *Colloque International Genre, Population et Développement en Afrique.*

64. **Sangho, O.** *et al.* **(2020)** "Comparaison des déterminants de l'accouchement à domicile dans deux quartiers en commune V de Bamako", p. 7.

65. **Sandberg, J.** *et al.* **(2012)** "Social learning about levels of perinatal and infant mortality in Niakhar, Senegal", *Social Networks*, 34(2), pp. 264-274.

66. **Sandberg, J.** *et al.* **(2019)** 'Social learning, influence, and ethnomedicine: Individual, neighborhood and social network influences on attachment to an ethnomedical cultural model in rural Senegal', *Social Science & Medicine*, 226, pp. 87-95.

67. **Sandberg, J.** *et al.* **(2020)** "A latent class analysis of attitudes concerning the acceptability of intimate partner violence in rural Senegal", *Population Health Metrics*, 18(1), p. 27.

68. **Sandberg, J.F.** *et al.* **(2021)** "Individual, Community, and Social Network Influences on Beliefs Concerning the Acceptability of Intimate Partner Violence in Rural Senegal", *Journal of Interpersonal Violence*, 36(11-12), pp. NP5610-NP5642.

69. **Sandberg, J.** *et al.* **(2012)** "Interpersonal Influence on Beliefs Concerning Childbirth Location in Rural Senegal: Evidence from the Niakhar Social Networks Pilot Survey", p. 30.

70. **Stephenson R., Ong Tsui A. [2002],** "Contextual influences on reproductive health service use in Uttar Pradesh, India", Studies in family planning, vol. 33

71. **SINGH P. K., KUMAR R. R., MANOJ A., Singh L. [2012],** "Determinants of maternity care services utilization among married adolescents in rural India", *PLoS One*, vol. 7, n° 2, pp. 1-14.

72. **Sala-Diakanda (1999),** Recherche des facteurs d'un recours de qualité aux soins pendant la grossesse, l'accouchement et le post-partum, Cas de la ville de Bafia, DESSD dissertation, University of Yaoundé II, IFORD, 96p.

73. **Soubeigua, D. (2005),** La continuité des soins obstétricaux au Burkina Faso: Niveaux et déterminants, DESSD dissertation, IFORD, University of Yaoundé II, 132p.

74. **Vallin J., Caselli J. and Surault P. (2002),** "Comportements, styles de vie et facteurs socioculturels de la mortalité" in Démographie : analyse et synthèse. Les Déterminants de la mortalité, vol. III edited by Graziella Caselli, Jacques Vallin and Guillaume Wunsch, Editions de l'INED pp: 255-305.

75. **Wilfried, M.G.-H., Jérôme, A.-N. and Valentin, E.K. (2018)** "Les Déterminants De L'accès Aux Services De Santé À Grand Bassam", *European Scientific Journal, ESJ*, 14(6), p. 124.

76. **Yanagisawa S., Sophal O., Ssusumu W. [2006],** "Determinants of skilled birth attendance in rural Cambodia", Tropical medicine and international health, vol. 11, no. 2, pp. 238-251

77. **Zoungrana (1993),** Déterminants socio-économiques de l'utilisation des services de santé maternelle et infantile à Bamako (Mali), PhD thesis in demography, Université de Montréal, 213p.

APPENDIX

Table 18: Rate of home births in the different districts of SSD - de Niakhar from 1983 to 2017.

NOMHAMEAU	NOMVILLAGE	Accouchem ent à domicile 1983_1987 en %	Accouchem ent à domicile 1988_1992 en %	Accouchem ent à domicile 1993_1997 en %	Accouchem ent à domicile 1998_2002 en %	Accouchem ent à domicile 2003_2007 en %	Accouchem ent à domicile 2013_2017 en %	Accouchem ent à domicile 2008_2012 en %	Accouchem ent à domicile 2017 en %
CENTRE	DAROU	88	100	57	80	58	12	68	0
CENTRE	DIOKOUL	89	89	92	81	81	55	71	50
CENTRE	KALOME NDOFANE	76	76	73	89	72	22	49	16
FELANE	KALOME NDOFANE	87	85	91	93	79	32	63	12
MBIND DIANA	KALOME NDOFANE	86	92	81	83	50	43	50	75
NDOFANE	KALOME NDOFANE	86	88	84	75	57	38	63	40
MBAFAYE	KALOME NDOFANE	86	100	60	43	88	0	25	0
NDIOBENE	KALOME NDOFANE	100	50	100	67	50	0	0	0
CENTRE	NGALAGNE	93	95	86	93	78	47	71	47

	KOP								
KHASSEME	NGALAGNE KOP	78	97	91	88	93	36	81	42
PIND TOK	NGALAGNE KOP	86	82	80	95	86	41	52	50
SOBEME	NGALAGNE KOP	89	100	88	94	94	53	69	33
GODAGUENE	NGALAGNE KOP	100	94	74	87	88	58	60	73
KHOUDOMBEDJ	NGALAGNE KOP	67	100	100	82	98	90	100	50
CENTRE	NGANE FISSEL	92	82	73	92	71	43	71	10
BADEME	NGANE FISSEL	85	72	91	93	53	43	71	60
TOK BILEB	NGANE FISSEL	82	88	76	83	64	33	32	33
HA PIND	NGANE FISSEL	85	100	88	83	63	50	45	33
CENTRE	NGAYOKHEME	25	100	100	0	33	0	0	0
MBIND PAMA	NGAYOKHEME	76	74	53	72	29	7	11	0
NDIOUDIOUF	NGAYOKHEME	87	88	87	100	87	35	76	33
NDIALO	NGAYOKHEME	85	93	80	86	74	54	68	55
MONEME	NGAYOKHEME	100	92	78	83	83	64	66	56
NGUILGANDANE	NGAYOKHEME	95	87	67	94	62	45	55	40

LEONA	NGAYOKHEME	81	85	79	89	78	23	23	18
NDIAYENE	NGAYOKHEME	67	59	61	89	67	9	32	23
MBONGAB	NGAYOKHEME	64	87	70	84	64	5	48	0
MBIND JAGA	NGAYOKHEME	100	100	83	94	100	30	50	100
CENTRE NGOTHIEME	SASS NDIAFADJI	94	93	88	81	60	30	34	29
NGODJILEME	SASS NDIAFADJI	92	86	80	78	67	47	51	38
NDIEDIENG NDIODIONE	SASS NDIAFADJI	93	92	88	86	62	32	70	25
NDOFENE	SASS NDIAFADJI	88	88	88	79	78	20	65	33
BAK MAK	SASS NDIAFADJI	100	100	100	100	67	25	67	0
MBIND BOURE	SASS NDIAFADJI	100	100	100	100	100	0	100	0
CENTRE	SOB	91	91	84	90	68	47	61	40
NDOFANE	SOB	94	94	80	87	83	42	70	0
PIND A KOP	SOB	82	100	100	100	70	60	80	0
BARY SINE	BARY NDONDOL	96	91	97	93	76	57	78	52
THIATHIAO	BARY NDONDOL	90	90	92	97	64	39	69	19

CENTRE	DATEL	100	94	95	90	94	66	86	25
DIAM DIOP	DATEL	100	96	100	87	91	62	78	40
PETHI GOR	DATEL	100	91	91	93	97	59	85	83
SOBEME	DATEL	95	100	97	94	88	79	90	69
TOK MBED	DATEL	100	100	95	100	87	90	91	80
NGANGAR	DATEL	70	100	92	73	92	40	56	33
SII MBONE	DATEL	100	100	86	94	81	77	90	75
CENTRE	LAMBANENE	87	80	97	90	85	33	78	32
DAME THIED	LAMBANENE	100	86	100	100	91	59	93	25
MBIND SIRA	LAMBANENE	88	73	94	89	67	60	80	50
NDIANGAYE	LAMBANENE	100	86	100	90	81	24	74	0
CENTRE	MBINONDAR	89	89	86	100	100	62	25	50
SINDIANKE	MBINONDAR	92	88	88	86	54	40	54	19
NDIALO DIALO	MBINONDAR	100	100	100	100	100	57	100	50
NGODJILEME	MBINONDAR	88	93	88	79	86	71	92	50
TOK NGOL	MBINONDAR	71	88	91	73	67	33	25	0
CENTRE	MBOYENE	92	92	98	89	76	87	84	67
MBOUR DIAK	MBOYENE	89	85	89	96	78	66	75	69
MBOYENE TOK	MBOYENE	100	88	86	88	88	79	86	88
CENTRE	NDOKH	88	77	95	91	64	47	69	64
NDOKH MBAD	NDOKH	92	94	93	83	72	66	85	64
PIND TOK	NDOKH	100	95	100	84	64	38	79	50

SANGAYE	NDOKH	85	75	93	93	92	53	57	40
CENTRE(PIND TOK)	NGANGARLAM	96	93	94	94	74	68	81	69
DOULEME	NGANGARLAM	95	94	90	91	83	62	82	30
KHAKHAL	NGANGARLAM	100	95	89	94	74	78	80	33
NDANG	NGANGARLAM	100	100	100	88	95	82	79	58
NDOFANE	NGANGARLAM	93	95	82	86	66	64	69	69
PETHI SAMBOUR	NGANGARLAM	89	95	95	94	64	45	77	71
PIND ALANG	NGANGARLAM	100	96	92	92	74	40	90	33
CENTRE	NGONINE	92	88	87	86	68	78	83	81
GADIAK	NGONINE	100	86	94	91	79	54	86	71
MBESS	NGONINE	92	90	89	89	81	58	73	68
SOBEME	NGONINE	89	95	96	92	80	59	69	29
DIFEME	POUDAYE	73	59	82	53	89	61	71	55
GATHI	POUDAYE	93	79	73	74	89	50	57	50
SASSAR MATOKHOL	POUDAYE	94	86	89	82	84	67	67	79
MBADATHIE	POUDAYE	53	85	86	96	83	80	69	33
NGAFOYE	POUDAYE	100	87	100	88	76	60	95	43
POUDAYE KAME	POUDAYE	93	80	67	91	70	86	62	100
SASSEME (TK)	POUDAYE	73	89	71	67	68	47	48	50
TOUNE	POUDAYE	89	95	70	86	89	71	79	50

CENTRE	TOUCAR	56	63	55	63	60	27	42	29
DAMONGAYE	TOUCAR	100	75	67	86	90	83	40	100
KAMA KAMA	TOUCAR	69	73	85	29	36	44	23	57
MBAP	TOUCAR	83	65	49	75	73	45	55	27
NDIAYENE	TOUCAR	76	69	76	75	76	38	64	43
NDIOUDIOUF	TOUCAR	87	88	93	82	84	53	61	33
NDOFANE	TOUCAR	75	71	25	69	79	47	43	100
NGANEME	TOUCAR	85	68	30	48	76	30	70	27
NGAOLEME	TOUCAR	53	65	82	59	77	33	31	0
NGOULANGUEM E	TOUCAR	73	69	90	83	88	67	50	100
NIOLELEME- NENEM	TOUCAR	38	43	78	88	86	10	43	0
PIND TOK	TOUCAR	88	61	50	81	76	37	72	30
SANGAYE	TOUCAR	88	90	89	61	86	33	54	25
SASSEME	TOUCAR	95	42	67	74	98	47	46	67
TOK ASSONE	TOUCAR	50	71	29	83	89	38	53	100
TOK NGOL	TOUCAR	88	76	57	64	82	43	68	50
TONGOGNE	TOUCAR	71	76	62	54	87	44	62	58
WAGBAKH	TOUCAR	86	77	74	84	95	60	61	44
DJAN MBAL	TOUCAR	100	85	91	83	100	53	31	100
CENTRE	DAME	92	88	90	83	81	56	56	50

CENTRE(MARON EME)	DIOHINE	76	67	62	53	50	28	44	19
BOTE	DIOHINE	88	86	67	45	48	23	51	0
MBELO NGUITHE	DIOHINE	87	89	85	71	83	77	71	38
NDIENE	DIOHINE	93	77	57	58	45	35	61	20
POULANDERE	DIOHINE	89	69	82	60	69	42	58	17
SASSAR	DIOHINE	93	82	56	33	34	15	26	7
SASSEME	DIOHINE	87	76	66	65	39	22	24	25
THITAR	DIOHINE	91	92	74	77	50	27	34	14
CENTRE	GADIAK	96	81	97	93	81	78	76	57
MEME A KOP	GADIAK	95	84	85	82	76	65	74	57
DAFEME	GADIAK	90	90	85	84	78	70	75	36
DIFNA MBELONGOUD	GADIAK	77	84	88	76	80	74	76	81
MBEDIED	GADIAK	86	91	93	84	81	79	95	80
MBELO NGUITHE	GADIAK	94	80	84	75	84	84	89	93
NGASSIAK	GADIAK	94	90	93	76	92	62	82	69
MBALAK-DAROU SALAM	GADIAK	100	90	82	89	80	82	79	80
SOUNDANE	GADIAK	100	100	91	86	94	81	55	92
BAK SEK	GADIAK	100	100	100	100	100	100	83	0
DAKH ROG	GADIAK	85	96	95	83	86	72	83	100

MBOBONGA	GADIAK	75	86	67	82	100	80	93	75
CENTRE	GODEL	100	98	90	87	84	64	78	45
BAK NDIEKE	GODEL	82	88	88	85	69	55	96	33
HABADA	GODEL	83	75	85	79	90	83	88	100
MBEL FATAR	GODEL	97	95	77	67	72	64	74	83
MBEL PIL	GODEL	100	87	96	93	93	76	85	100
NENE KOR	GODEL	88	94	94	84	95	74	80	56
NGUERANE	GODEL	100	88	97	100	70	64	74	78
BAK WAGANE	KHASSOUS	86	95	85	81	82	23	68	0
KHASSOUS-DIEGNAK	KHASSOUS	100	73	76	75	57	63	71	75
KHASSOUS-THIEDO	KHASSOUS	95	86	82	83	79	73	70	82
MOURTALE	KHASSOUS	90	82	92	73	79	43	76	25
NGUELO-AMAK	KHASSOUS	93	90	93	94	88	55	86	56
NGUELO-ATEB	KHASSOUS	89	88	91	89	95	83	82	86
CENTRE	KOTHIOKH	91	98	87	78	77	68	77	64
BAK-MBANE	KOTHIOKH	95	93	91	88	66	47	72	40
NDOFO-MBOUDJ	KOTHIOKH	95	88	83	90	73	62	71	50
NENE-KOR	KOTHIOKH	88	93	93	91	97	57	72	50
NGUITH	KOTHIOKH	95	100	86	81	86	53	71	75
KALALE	KOTHIOKH	100	100	88	100	81	60	72	67

CENTRE	LEME	94	91	88	69	73	39	53	40
CENTRE(SINDIANE)	LOGDIR	97	95	83	78	57	19	38	12
MBOUMI	LOGDIR	95	83	94	68	86	51	66	50
MONE	LOGDIR	92	95	82	58	75	21	63	0
SESSENE	LOGDIR	97	85	93	75	74	44	47	67
SOUBEME 1	LOGDIR	97	92	96	72	66	50	86	0
SOUBEME 2	LOGDIR	93	100	95	94	74	52	32	100
CENTRE	MEME	83	92	88	82	29	14	30	33
NDADAF	MEME	80	100	100	100	50	0	33	100
SOUBEME	MEME	100	100	80	70	70	23	33	0
CENTRE	MOKANE NGOUYE	93	80	76	88	79	39	41	33
NGUESSIAM	MOKANE NGOUYE	100	82	100	100	50	50	38	50
NDOUTKI	MOKANE NGOUYE	94	95	76	94	79	40	53	33
NGOUYE	MOKANE NGOUYE	100	88	50	88	86	22	88	50
TOUCARKE	MOKANE NGOUYE	86	95	90	100	67	55	39	25
CENTRE(SASSEME)	NGARDIAME	95	89	68	65	64	48	62	25

KAMSATE	NGARDIAME	94	97	86	100	81	56	62	33
NGANEME	NGARDIAME	88	89	100	91	82	68	85	75
PIND TOK	NGARDIAME	96	75	95	70	79	43	79	0
TOK AKAL	NGARDIAME	90	94	71	88	71	53	72	0
CENTRE(MBAFA YE)	POULTOK DIOHINE	100	92	95	86	91	43	67	25
MBALEME	POULTOK DIOHINE	100	94	91	74	47	50	67	50
NDIODIONE	POULTOK DIOHINE	100	79	89	73	71	50	54	0
NGODJILEME	POULTOK DIOHINE	100	93	90	88	88	40	75	0
SESSENE	POULTOK DIOHINE	93	88	88	76	70	42	60	28
TOK GOL	POULTOK DIOHINE	90	88	92	79	85	51	64	56

Table 19: Description of non-use of medical assistance in childbirth according to the woman's marital status.

Marital status	health center	concession	Total	% home birth	% assisted delivery
Single	493	125	618	20	80
divorced	5	1	6	17	83
bride	2358	1177	3535	33	67
Total	2856	1303	4159	31	69
Chi-square probability		Pearson chi2(2) = 42.3625 Pr = 0.000			

Table 20: Description of non-use of medical assistance in childbirth by number of prenatal consultations.

number of CPN	health center	Concession	Total	% home birth	% assisted delivery
[Less than or equal to 2	438	525	963	55	45
[Greater than or equal to 3	2968	991	3959	25	75
Total	3406	1516	4922	31	69

Chi-square probability	Pearson chi2(1) = 315.9572 Pr = 0.000

Data sources : Niakhar database and author's calculations.

Table 21: Description of non-use of medical assistance in childbirth by age of woman.

Age group	health center	concession	Total	% home birth	% assisted delivery
[<=20]	483	107	590	18	82
[20-34]	2223	996	3219	31	69
[35-49]	687	407	1094	37	63
Total	3393	1510	4903	31	69
Chi-square probability	Pearson chi2(2) = 65.4747 Pr = 0.000				

Data sources : Niakhar database and author's calculations.

Table 22: Description of non-use of medical assistance in childbirth, by parity achieved.

Parity Achieved	health center	Concession	Total	% home birth	%assisted delivery
1	566	123	689	18	82
[2-3]	818	360	1178	31	69
[4 and +]	1747	988	2735	36	64

Total	3131	1471	4602	32	68
Chi-square probability	Pearson chi2(2) = 85.9300 Pr = 0.000				

Data sources : Niakhar database and author's calculations.

Table 23: Description of non-use of medical assistance in childbirth, by economic situation.

Economic situation	health center	Concession	Total	% home birth	% assisted delivery
Poor	2270	1165	3435	34	66
Medium	479	140	619	23	77
Rich	657	211	868	24	76
Total	3406	1516	4922	31	69
Chi-square probability	Pearson chi2(2) = 52.2502 Pr = 0.000				

Data sources : Niakhar database and author's calculations.

Table 24: Description of non-use of medical assistance in childbirth, by woman's level of education.

Education level	health center	Concession	Total	% home birth	% assisted delivery
primary	569	212	781	27	73
secondary means	464	73	537	14	86
Koranic	112	51	163	31	69
university	14	3	17	18	82
Out of school	2238	1177	3415	34	66
Total	3406	1516	4922	31	69
Chi-square probability	Pearson chi2(8) = 111.7378 Pr = 0.000				

Data sources : Niakhar database and author's calculations.

Table 25: Description of non-use of medical assistance in childbirth, by woman's caste.

Caste	health center	Concession	Total	% home birth	% assisted delivery
Farmers	2829	1314	4143	32	68
DB Tiedo	193	42	235	18	82
dep DB	118	60	178	34	66
Griots	138	63	201	31	69
Artisans	20	6	26	23	77

not tight	30	5	35	14	86
unknown	78	26	104	25	75
Total	3406	1516	4922	31	69
Chi-square probability	Pearson chi2(6) = 27.6397 Pr = 0.000				

Data sources : Niakhar database and author's calculations.

Printed by Books on Demand GmbH, Norderstedt / Germany